MAYA WISDOM AND THE SURVIVAL OF OUR PLANET

MAYA WISDOM AND THE SURVIVAL OF OUR PLANET

Lisa J. Lucero

OXFORD
UNIVERSITY PRESS

OXFORD
UNIVERSITY PRESS

Oxford University Press is a department of the University of Oxford.
It furthers the University's objective of excellence in research, scholarship,
and education by publishing worldwide. Oxford is a registered trade mark of
Oxford University Press in the UK and in certain other countries.

Published in the United States of America by Oxford University Press
198 Madison Avenue, New York, NY 10016, United States of America.

© Oxford University Press 2025

All rights reserved. No part of this publication may be reproduced, stored in a retrieval system, transmitted, used for text and data mining, or used for training artificial intelligence, in any form or by any means, without the prior permission in writing of Oxford University Press, or as expressly permitted by law, by license or under terms agreed with the appropriate reprographics rights organization. Inquiries concerning reproduction outside the scope of the above should be sent to the Rights Department, Oxford University Press, at the address above.

You must not circulate this work in any other form
and you must impose this same condition on any acquirer

Library of Congress Cataloging-in-Publication Data
Names: Lucero, Lisa Joyce, 1962– author.
Title: Maya wisdom and the survival of our planet / Lisa J. Lucero.
Description: New York, NY : Oxford University Press, [2025] | Includes bibliographical references and index. | Summary: "Maya Wisdom and the Survival of Our Planet presents the Maya way of seeing and interacting with the world that embodies lessons and provides solutions to ensure a sustainable future of Earth. This book is based on over three decades of working with Maya associates in Belize, Central America on the ancestral Maya as an archaeologist and approaches the future through the lens of the Maya non-anthropocentric inclusive worldview. Ancestral Maya people worked with, not against, nature. Nor did they privilege humans at the expense of nonhumans. Their engagement with the tropical environment was expressed in a landscape of green cities, farmsteads, gardens, fields, forests, and sacred places. The Maya built green cities that drew people in through royal reservoirs, a system that lasted over 1,000 years in the southern lowlands (c. 300 BCE–900 CE). After taking the reader on a journey through Maya history, their tropical world, and how they lived in it and engaged with nonhumans through ceremonies, the book concludes with concrete solutions that bridge the past and present for the future. Conditions are not going to change but people can. Maya resilience is a testament for how to move forward, and this book provides a roadmap on how to do so"—Provided by publisher.
Identifiers: LCCN 2024030448 (print) | LCCN 2024030449 (ebook) |
ISBN 9780197765708 (hardback) | ISBN 9780197765722 (epub) | ISBN 9780197765739
Subjects: LCSH: Mayas—Ethnobiology—Belize. | Maya philosophy—Belize. |
Mayas—Belize—Social life and customs. | Traditional ecological knowledge—Belize. | Climate change mitigation—Belize. |
Environmental archaeology—Belize. | Belize—Antiquities.
Classification: LCC F1435.3.E73 L84 2025 (print) |
LCC F1435.3.E73 (ebook) | DDC 972.81/016—dc23/eng/20241029
LC record available at https://lccn.loc.gov/2024030448
LC ebook record available at https://lccn.loc.gov/2024030449

DOI: 10.1093/oso/9780197765708.001.0001

Integrated Books International, United States of America

To Our Family

PREFACE

We now live in the Anthropocene, an age of our own making. We have altered the Earth's atmosphere, its bodies of water, and its landscapes. Fossil fuels and livestock emit greenhouse gases (CO_2 and methane especially) that envelope the Earth in a blanket, warming the planet enough to change weather patterns, melt glaciers, raise sea levels, and generally wreak havoc. We also live in a world where resource depletion and deforestation continue to detrimentally alter the Earth. There is no Planet B, as United Nations Secretary-General Ban Ki-moon stated while addressing the world in 2018.[1] Many of us look to governments to deal with these issues. I look to the Maya, past and present. They appreciate the fact that we need the Earth to survive—but that it does not need us. The Earth and its inhabitants will thrive and continue without humans. Thus, our continued existence on this planet demands that we think and act differently. The Maya lived sustainably for millennia. Excavations of their homes indicate that they housed generations for over 1,000 years, a feat only possible through careful collaboration and coexistence with soils, water, and forests and their animal inhabitants. The survival of our planet may well hinge on practices like those of the Maya.

Maya Wisdom and the Survival of Our Planet is a cumulation of decades working with Maya studying the ancestral Maya. It approaches our future through the lens of the traditional Maya worldview, where people worked with, not against, nature. My goal in this book is to open your eyes to the traditional Maya way of seeing and interacting with the world to provide solutions to ensure a sustainable future for our planet.

Like everyone else in Western society, I view the world through an anthropocentric lens where living and nonliving nonhumans such as forests, birds, sky, clouds, insects, mammals, chert, clay, and other resources, are at our disposal to ensure our survival. While our goal is laudable, our approach is not. We use any means possible to obtain what we need, typically at the expense of the nonhuman world, including deforestation, overuse of resources, and the extinction of, for example, animals and pollinating insects.[2] In fact, our anthropocentric worldview embeds most ideas to sustainably address climate change because they privilege humans. But we can't do it alone. To save ourselves, we need to save nonhumans, too.

Maya wisdom is the main theme of this book. It is relevant in today's changing and complex global world because it is rooted in how people engage with rather than dominate their environs. Why? Because conditions are not going to change—but *we* can. I offer solutions (Chapter 10) on how we can divert from our current unsustainable path based on insights I have learned from both the ancestral and living Maya over the course of my 35 years conducting archaeology in Belize (Chapters 1–9). The archaeological record, historical accounts, ethnographies, Maya stories and poems, and conversations with Maya foremen and excavation assistants, some of whom have worked with me for over 25 years, as well as with their families, have opened my eyes to a different way of seeing and interacting with the nonhuman world.

Some solutions I suggest are outside the box—really outside the box. For instance, for those of you who own a swimming pool, I hope to convince you to turn it into a self-cleaning water garden like ancient Maya urban reservoirs. You can drink the water, harvest fish and aquatic plants, attract waterfowl—and swim!

To appreciate Maya wisdom, it is first necessary to be aware of where it emerged and how it shaped how the Maya lived for thousands of years in the tropical environment through knowledge that has been passed down from elders and parents to children from generation to generation. I bridge the past and present for the future benefit of our planet and all who live on it.

I share what's it like excavating and exploring in the jungle with Maya associates as well as knowledge they and the archaeology record reveal. In fact, I could not have written this book without being an archaeologist. That said, I did not expect to write about the Maya inclusive worldview—I started out interested in how political leaders acquire power over others, particularly via ritual, which soon led to my interest in how climate change intersected with Maya history, especially kingship, which then led to my interest in ancient Maya water management systems, specifically their construction of self-cleaning reservoirs. These topics highlight Maya adaptive skills and sustainable practices.

***

On my first visit to Belize in 1988 as an archaeology graduate student at UCLA, a farmer from the modernized Mennonite town of Spanish Lookout said to me, "Da Maya, dey smarter dan we" in Kriol, the local Creole spoken by most Belizeans. We were standing in a treeless landscape surrounded by the remnants of the ancient Maya city of Barton Ramie along the Belize River made famous in the 1950s by Gordon Willey of Harvard University as part of one of the first archaeology settlement studies. At the time, I was part of a salvage archaeology project directed by Anabel Ford. Farmers were clearing and plowing the entire urban core and bulldozing pyramid temples to use as fill. Our goal was to collect as much information as possible before plowing and bulldozing destroyed everything.

The Maya deserted this city and the surrounding area in the early 1500s CE for reasons that are still unknown. What we saw on the surface was only the tip of the iceberg. It was the most recent evidence of over 2,000 years of Maya urban building and living. Mennonites, who came to Belize beginning in the 1950s, only buy or lease land with Maya mounds (collapsed structures) because they know the ancestral Maya lived on or near fertile land. This does not stop them from plowing basically everything in sight—forests and mounds alike. All they care about is farming. The issue is the short-term strategies upon which they rely: extensive deforestation, monocropping, and the use of chemical fertilizers, herbicides, and pesticides that leave behind a bleak landscape devoid of trees, bushes, birds, mammals, and reptiles.

In stark contrast, Classic Maya (c. 250–900 CE) cities surrounded by rural farmsteads encompassed home gardens, orchards with a diverse array of tree species, urban fields, self-cleaning reservoirs, carved stone monuments, ball courts, houses of all sizes, monumental temples and palaces filled with beautifully painted ceramics, and obsidian and jade items inscribed with exquisite hieroglyphs. And the Maya created and built all this without beasts of burden or metal tools. They relied on their ingenuity, stone tools, and an endless green energy source—human labor.

Because Maya cities in the southern lowlands in present day Belize, northern Guatemala, southeastern Mexico, and western Honduras have been unoccupied for over a thousand years since c. 900 CE, most of their original names have been lost to history; we now know them as Barton Ramie, Tikal, Calakmul, Caracol, Naranjo Sa'aal, and other names. Archaeologists often named cities after logging camps. The camps, you see, are located at *aguadas*, that is, natural depressions the ancestral Maya lined with clay to contain water, and Classic Maya urban reservoirs. They are often the only reliable water sources during the long dry season when loggers do their thing. For instance, "Tikal" is from the Yucatec Mayan *ti ak'al*, which translates as "at the water hole."[3] Epigraphers have found and deciphered the names of a few cities, including Palenque (Lakamha' or "Big Water"), Tikal (Yax Mutal or "First/Green-Blue Bundle"), and Lamanai ("Submerged Crocodile").

Early archaeologists were quite colonial in their attitude and approach—for example, only collecting the "pretty" stuff and throwing the rest away and taking the "best" stuff for their museums back home. They also discriminated against the local Indigenous peoples. Things are much better now, for the most part.

The Belize Government owns all Maya sites in Belize. Landowners, most of whom are not Maya, are given rights as curators of artifacts they find on their property. If they keep artifacts, for example, from plowed fields, they need to register them through the Belize Institute of Archaeology by listing, describing, and photographing them. Not many do. In Spanish Lookout I've seen a *metate* (a rectangular granite surface with a concave depression for grinding maize) being used as a dog bowl and another one placed on a pedestal as a bird bath. Archaeologists often register collections on landowners' behalf, as we did for Banana Bank Lodge's owners (where we often stay while in Belize). The owners of Banana Bank, John and Carolyn Carr, farm over 800 hectares (2,000 acres) and have amassed a large collection of household items from plowed fields—*manos* (handheld granite shaped stones) and *metates* (used together to grind maize), stone implements, shell ornaments, obsidian, and ceramic vessels and sherds. Their collection serves as teaching tools for Belizeans, students, and tourists.

***

There are two things you need to know about archaeology. First, it is inherently destructive. Once we remove a stone brick from a wall, excavate a burial, dig through a floor, collect an assemblage of arranged items, we can't put anything back in its exact placement. So, before we remove anything, we document everything—draw, photograph, map, and describe. Second, it is very slow going. But the information we collect is priceless. And there is nothing like excavating ancient Maya buildings or finding artifacts—small obsidian or jade fragments, large sherds, chert hoes and blades, *manos* and *metates*, shell, animal bones, and other items. You don't need to find a tomb to get excited about archaeology.

The time is right for this book, not only because we live in a world where we all will have to adapt to an increasingly warming planet, but because the archaeological record in many parts of the world is being lost to looting, urban sprawl, construction projects, and destruction for the sake of destruction.

The current state of the world is a call to arms. This book is my response. Explorer and scientist Alexander von Humboldt (1769–1859) stated 175 years ago that the world comprises an interwoven web of life like a "thousand threads." We can repair what has been torn asunder. There is hope for our future. We can change how we see the world. We can change how we act. This book provides a roadmap for how to do so.

ACKNOWLEDGMENTS—35 YEARS' WORTH

First and foremost, I want to acknowledge and thank my Maya friends and associates. My field work in Belize and this book would not have been possible without their support, assistance, stories, and experiences. As a graduate student, I worked with men from Bullet Tree Falls and Santa Familia: Theophilus (Teo), Gonzolo, and Alberto Williams; Narciso Torres; Gustavo (Gus) Manzanero; Julio and Armando Martinez; and Beto, Amalio, Umberto, Nolberto, Javier Bacab, and Jerry. Since starting my own project, I have worked with people from the Valley of Peace Village: Zedekiah Scott (Mr. Scott); Cleofo, Stanley, and Mark Choc; José Ernesto, José, Carlos, and Javier Vasquez; Juan Antonio Lópes; Vicente Cal; Besi Alvarez (Rodriguez); Isabel Ascencio (Don Luna); Rutilio (Tilo) and Rejolio (Antonio) Luna; Joel and Jeremias Portillo; Rene Penido; Rafilo Sansores; Julio Rodriguez; Rafeal Magana; Rene Penido; Henry de Paz from Buena Vista; Marcial, Javier and Alex Arteaga; Alejandro (Javier) Gil; Ismael Blandon; Mario Rivera, Wilman Mendez; Valdemar Vasquez; and Yoel Ramos.

Their families have been just as important for their friendship and wonderful meals: Mrs. Williams and Mrs. Torres when I was a graduate student (and Bob and Alice Hurley), and Mrs. Scott when I started my own project in 1997. This task was later taken on by the Martinez family when I was working at Saturday Creek in 2001 and then by Miss Louisa (Cleofo's wife) and later her daughters Lucy and Jennifer and daughter-in-law Nala. Ernesto's daughter Jovelina has been instrumental in helping me arrange things in Belize before I arrive.

I have also relied on the support of the Belize Department of Archaeology, later the Belize Institute of Archaeology of the National Institute of Culture and History, which was led by Harriot Topsey and John Morris when I was a graduate student, later by Dr. Allan Moore, Dr. Jaime Awe, Dr. John Morris, and currently by Dr. Melissa Badillo. The support of Institute of Archaeology archaeologists over the years has been invaluable—Brian Woodye, George Thompson, Dr. Rafael Guerra, Josue Ramos, Paul Smith, David Griffith, Antonio Beardall, and Sylvia Batty.

Landowners have been quite generous about allowing us to work on their property: Banana Bank Lodge (John and Carolyn Carr, Leisa Carr-Caceres, and their wonderful staff), Yalbac Ranch/Forestland Group (Mike Hincher, Hunter Jenkins, and Alex Finkral), and the Spanish Lookout Community Corporation. The manager of Yalbac Ranch, Jeff Roberson, was incredibly supportive of our research, as were the landowners, Forestland Group. He loaned us satellite radios, cleared roads for us, and allowed me to store my 20-foot container at their sawmill. His passing early in 2022 was a major loss for all of us. The guard at the South Gate, Ernesto Velasquez (now the Head Ranger for Belize Maya Forest Trust) was the best, as are the rest of the Yalbac Ranch team. Nathan Jaeger, whose family co-owns Banana Bank, has been an incredible help, especially with our field vehicles.

In San Ignacio (Cayo) where I spent my formative years, Bob and Nettie Jones and their family, Paulita Bedran Figueroa of the San Ignacio Hotel and her family, John and Mariam Roberson, Yvette Smith, and others became close friends. As a graduate student under the excellent and challenging tutelage of Anabel Ford, I worked with some great volunteers from the University Research Expeditions Program—David Brennan and Robert Vitolo in particular, as well as

UCLA Friends of Archaeology volunteers—Berniece Skinner, Lady Harrington, and Jean Wood.

Since becoming a professor, first at New Mexico State University (1997–2007) and then at the University of Illinois at Urbana-Champaign (2007–present), I've been lucky to have some great graduate students and colleagues in the field. My first field director, Andrew Kinkella (PhD, 2009), was a lifesaver during those early years of my project. Later graduate crew chiefs and field directors have been just as vital—Colleen Lindsay, Jessica Harrison, Erin Benson, Aimée Carbaugh, Jean T. Larmon, Rachel Taylor, and my current PhD student, Yifan Wang. I have run several field schools over the years and taken undergraduate volunteers—they have been so helpful. And a special shout-out to undergraduate Katherine Mass, who wrote an excellent honor's paper (2012) comparing lotus and water lily flowers.

I worked with some great colleagues in the field, for which I am thankful, Kirsten Olson in particular. Later, there were surveyors James Wakeman and Lonnie Mehlin, electrical engineer Lonnie Ludeman, historian William Poe, and soil scientist Susan Hayes. I have been lucky to work with some amazing ceramic specialists: James Conlon, Jennifer Ehret, and Laura Kosakowsky. And there are too many fellow archaeologists to mention—you know who you are.

Since I started the diving expeditions at Cara Blanca, I have been introduced to an entirely different world of amazing experts: Patricia Beddows, a hydrologist and geochemist at Northwestern University; underwater videographer Marty O'Farrell; cave exploration divers Robbie Schmittner, Kim Davidsson, and Bil Phillips; and underwater archaeologist Andrew Kinkella. Bob Hemm of the Explorers Club documented the first diving expedition in 2010. Later diving expeditions included the owner of Belize Diving Services, Chip Petersen, nature photographer Tony Rath, and Patrick Widmann. Retired Bureau of Land Management paleontologist H. Gregory McDonald and dendro-climatologist and tropical tree specialist Brendan Buckley of the Lamont-Doherty Earth Observatory at Columbia

University collected fossils and tree specimens. To collect sediment cores from pools, Chip had help from oceanic engineer Anthony Tedeschi and dive instructor Peter Goebels. Videographer Thomas Hines joined us to document our explorations at pools and other unexplored Cara Blanca realms. Finally, I owe a big thank you to Thomas Franklin of the U.S. Forest Service for his eDNA expertise and rock-climbing skills at the Motmot Sinkhole.

Vilma Fialko, director of the Naranjo-Yaxha project in Guatemala, gave amazing tours of the massive Maya city of Naranjo (more properly, Naranjo Sa'aal) in 2007 and 2018. And I got some great insights about ancient Maya cities from the codirector of the Greater Angkor Project, Roland Fletcher, and Christophe Pottier of the École Française d'Extrême-Orient, director of the Cambodian-French Archaeological Mission on the Angkor Region, and a codirector of the Greater Angkor Project. Input from archaeology colleagues and friends Timothy Pauketat and Susan Alt firmed up my plans to start the rescue project during their 2015 visit.

I could not have done any field work without financial support, for which I am grateful: UCLA Friends of Archaeology, Sigma Xi, I.I.E. Fulbright, New Mexico State University Summer Research Awards, University of Illinois Research Board and Center for Latin American and Caribbean Studies, National Geographic Society (no. 8673–09), National Science Foundation (BNS no. 9111535, SBE-BCS 0004410, SBE-BCS 1110005, SBE-BCS 1249235, and BCS 2020465), and donations from David Brennan, Dr. Robert Vannix, Berniece Skinner, Robert Vitolo, and the Forestland Group.

I wrote this book as part of a fabulous writing group—thank you, Ellen Moodie and Jeannie Larmon for your great feedback. And thanks to those who read bits and pieces of this book or talked through ideas with me: Chip Colwell, Nadia Durrani, Timothy Earle, Brian Fagan, Eugene Linden, Gabrielle (Gabs) Lucero, Timothy Pauketat, Andrea Pitzer, Patrick Roberts, Vernon Scarborough, Monica Smith, and Dan Vergano. And I want to thank Stefan Vranka, executive editor at Oxford University Press, for his excellent advice, support, and feedback, all of which made this a better and more relevant book.

Finally, I want to thank my family, especially my sister and brother, Laura and Larry Lucero. Five of my nine nieces and nephews have visited me in Belize: Karman and Gabs Lucero (Larry's), and Audrey, Anthony, and Andrea Filson (Laura's). Though none of them have become archaeologists, they all had a great experience. Now if only I could get their parents to visit me (I'm talking to you, Laura, Larry, and Deb).

In Memoriam

Margarita (2007), Jennifer Ehret (2008), Mr. Scott (2011), Don Luna (2011), Miss Louisa (2017), Bil Phillips (2017), Bob Jones (2019), Peter Jones (2021), Jeff Roberson (2022), George Thompson (2023), and finally, my parents—Jim (2015) and Ida Lucero (2018).

CONTENTS

Preface | vii
Acknowledgments—35 Years' Worth | xiii

Introduction | 1

1. Setting the Stage: Then and Now | 14

2. The Maya | 30

3. Chahk, the Capricious Rain God | 47

4. The Maya Inclusive Worldview | 63

5. Relations with the Three Realms | 83

6. The Maize People | 105

7. House and Cosmos | 121

8. Water Lily Kings | 144

9. *Yax* Cities | 158

10. The Survival of Our Planet | 175

ENDNOTES 205
FURTHER READINGS 227
INDEX 233

Introduction

"You mean near the three palm trees in a row?" asked one of my foremen, Cleofo Choc.[1] I nodded, amazed as usual because all I saw in the middle of the jungle in central Belize was green everywhere—a blur of trees, shrubs, and vines. But Cleofo, a Mopan Maya, knew exactly where we had excavated the previous season. It had been 10 months since we had last been in this area, which is a long time in the tropical forest; vegetation quickly takes over, shrouding the ancient Maya house we had excavated. He sees much more than we do, especially compared to someone like me, who grew up in American suburbia. In Cleofo's eyes and mindset, each bush, tree, and vine is unique and has a purpose. He learned about the tropical forest and its myriad of denizens from his father, and he has passed on this knowledge to his sons. This evolving chain of knowledge continues as it has for thousands of years, despite the last five centuries of conflict since the arrival of Europeans and subsequent forced nucleation into villages (*congregación*), exposure to epidemic diseases, land-grabbing, disenfranchisement, civil wars beginning in 1960 and lasting through the mid-1990s, and increasing deforestation and changing climate.

Maya knowledge passed down through the millennia has allowed them to survive, but not at the expense of the environment and its nonhuman inhabitants. This knowledge, along with their nonanthropocentric inclusive worldview, shapes how the Maya, then and now, interact with the forest and its inhabitants via culling and promoting certain species of flora and fauna, clearing land, setting intentional fires for hunting, gathering wood for fuel, and extracting resources such as clay and chert for pottery and tool production, as well as limestone and wood for construction materials. There is so much value in the knowledge they have. It is part of who they are and is embodied in their daily existence.

The Central American tropical ecosystem is a pulsating, breathing entity. It is a world with noticeable seasonality, high biodiversity, hurricanes, droughts, endemic diseases, and so much more. And lots of spiders and insects. Ancestral Maya engagement with the tropical environment was expressed in a landscape of green cities, farmsteads (including home gardens) and fields, forests, and untouched sacred places. Neither the Maya nor the environment over-taxed each other. They coexisted, indicated by the fact that the ancestral Maya did not cause the extinction of any flora or fauna before the arrival of the Spanish.[2] This coexistence stands in stark contrast to our world today, where about 73 genera have become extinct since 1500 CE.[3]

The Maya have adapted to many changes over the millennia—demands from Maya kings, Spanish and English colonial rule, and massive droughts. And they continue to adapt today in Central America and elsewhere, and many of their ancestral practices are alive and well. Their resilience is a testament for the rest of us.

This is what the living and ancestral Maya have taught me: we are all connected. And that includes nonhumans. We are part of one large family and need to take care of one another. The ancestral Maya acknowledged nonhuman kin and didn't need genetics to show this connection. They engaged the world in a manner that did not privilege humans, thus sustaining them and their environment. They *lived* the circle of life.

Family was and is the unit of action. The family or household is the basic building block of society—socially, economically, and politically—and of survival and change. Together, building blocks can move mountains. Family is what you make of it—partners, spouses, roommates, individuals, nuclear and extended families, or whatever you define as your family. I use "family" to represent the household and unit of action.

Family action is only the beginning. The bottom-up insights I detail in this book, together with top-down mitigation and action by governments and transnational corporations, can make a difference for our future and that of our planet. Naturalist-philosopher Aldo Leopold provides an excellent example of how this process works, showing how a group of women worked with the U.S. government to save wild birds being hunted to extinction in the early twentieth century for their exotic plumage, which was used to adorn the heads of the very same women who fought against the practice,[4] demonstrating how people and governments can work together to instill change. In contrast, when the United States banned liquor consumption during Prohibition (1919–1933), it failed because people didn't support it (i.e., bottom-up support). Speakeasys emerged all over the United States. It also had the unintended consequence of expanding the reach and power of organized crime.[5]

Today, it will take all of us to change our current path, which is vital for the survival of our planet. Small to large actions add up and can benefit everyone.

Through the lens of the Maya, I show how we can change by engaging differently with our environs and addressing these questions: How did the Maya farm for 4,000 years without destroying their environment? How did they feed more people in the past than are living in the same area today? How did they survive in the humid jungles of Central America, building not only homes and garden plots, but also the great Maya cities?

Answers to these questions are relevant today and are found throughout this book. They include relying more on diverse and flexible long-term subsistence strategies, adopting a more inclusive or nonanthropocentric worldview that does not privilege humans, and appreciating the benefits of traditional Indigenous knowledge. The Maya show that adaptation and change begin in the home and that it is crucial to learn from and collaborate and coexist with nonhumans. We can learn from the ancestral Maya how, for example, to transform our cities into green cities that rely on local networks incorporating rural communities. After all, some of their cities, which they built with stone tools and hard work, lasted over 2,000 years.

The Divorce of Nature and Culture

The Renaissance (thirteenth to seventeenth centuries) in Europe was a time of exploration and "discovery"—from a Western perspective, anyway.[6] Christopher Columbus first set foot on Caribbean shores in 1492. This era also saw the increasing specialization of disciplines and advances in mathematics, physics, astronomy, biology, chemistry, and zoology. Martin Luther (1483–1546), born into a merchant family, initiated the Reformation, which not only rocked the foundation of the Catholic Church, but also came to embrace what Max Weber referred to as *The Protestant Ethic and the Spirit of Capitalism* (1905). This new kind of Christianity and its ideas espoused a religious construct that crystalized the anthropocentric hierarchy where nonhumans existed to benefit humankind. "In striving for the kingdom of God, the *kingdom* of Earth was forgotten."[7] This way of thinking and living differs from, for example, that of Native Americans:

> Both the story of Genesis and the story of Sky Woman tell of a world that existed before humans. Both tell of a woman and nonhumans interacting to create significant changes in creation as well as for humankind. In the latter, the relationship between animals and this female is regarded as sacred and ritualized over generations. This relationship also becomes the foundation for future clan

systems, ethics, governance, ceremonies, etc. In the former, the female becomes responsible for all the pain of childbirth and resentment for being cast out of paradise. The interaction of Eve and the Serpent results in shame and excommunication from nature. Additionally, future dialogue and communication with animals becomes taboo and a source of witchcraft. It is at this point of conflict where thought, perception, and action are separated from the supposed inertia of nature.[8]

In the Christian origin history, however,

> we see how the interaction between a female (Eve) and nonhumans (Serpent, Tree of knowledge, apple) led to the damnation of all future humankind…It also meant that the garden, in which they were able to reside, quickly became a place where humans were cast out. They were no longer *of* their surroundings, but outside of them.[9]

The Enlightenment (seventeenth to eighteenth centuries) is synonymous with the scientific revolution, set forth by Francis Bacon (1561–1626). At its core is reductionism, where complex systems or phenomena are broken down into parts and studied to better understand the whole. Great thinkers emerged whose ideas about the separation of nature and humans set the stage for the Anthropocene, where the overuse of resources, pollution of Earth, emission of greenhouse gases, and deforestation have created a hotter atmosphere and scarred landscape.

René Descartes (1596–1650) espoused the concept of dualism, where humans are viewed as distinct and separate from nature. Isaac Newton (1642–1727), famous for his discovery of the laws of gravity and motion, posited that all aspects of nature could be understood by laws or mechanical processes, further separating nature from culture and humans. This intellectual and cosmological turn of events resulted in the notion that nature serves humans. Humans thus can exploit nature. The divorce of nature from culture has had major repercussions, epitomized in capitalism, where nature is valued for profit margins. Yet we cannot exist without trees, soil, water, rain, flora, fauna, and the other nonhumans of the world. We rely on the Earth's bounty to survive. We need it. The Earth, however, does not need us. There is no Planet B.

The ancestral Maya collaborated with "nature" (they don't have a word for "nature," as I explain in Chapter 4), resulting in millennia of sustainable living, conveyed on the right side of Figure 1 (no pun intended). The lack of a culture-nature dichotomy—that is, no hierarchy—results in a sustainable existence without privileging humans because decisions and actions consider nonhumans. Our own history is one of trying to control or tame nature, resulting in the Anthropocene on the left side of Figure 1. This hierarchy also involved Euro-American dominance of Indigenous populations like the Maya, where inclusive worldviews have been suppressed due to imperialism and colonialism. It is hierarchy versus a world where everyone is on the same plane and connected, like a tree and its trunk, with numerous branches of all sizes.[10]

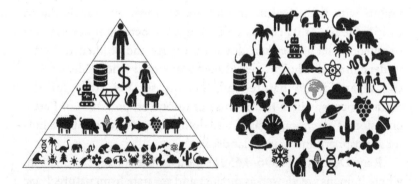

FIGURE 1 *Anthropocentric versus inclusive worldview*

Source: Modified from Figure 4 in L. J. Lucero and J. Gonzalez Cruz. 2020. Reconceptualizing Urbanism: Insights from Maya Cosmology. *Frontiers in Sustainable Cities: Urban Resource Management* 2:1. doi: 10.3389/frsc.2020.00001. Courtesy of the Creative Commons Attribution License (CC BY): https://creativecommons.org/licenses/by/4.0/

We created the rupture between nature and culture, and we can reverse this course by breaking down the hierarchy and striving for a nonanthropocentric lifestyle. We can put ourselves back into the circle of life instead of trying to lord over it. The core of how we can rethink the nature-culture dichotomy lies in the following chapters.

We can change our path and move forward using Maya wisdom. We can't go back to the past, of course. But it is possible to redirect our interaction with the nonhuman world using ideas *from* the past—with the caveat that what has been sustainable for millennia may not be flexible enough to continue or may need re-evaluation in view of global climate change. As an example, currently, the tropical belt is geographically defined as 23.5° latitude above (Tropic of Cancer) and below (Tropic of Capricorn) the equator. It rains daily in latitudes between 5° above and below the equator.[11] Between 5° and 23.5° latitude we find seasonal dry and wet periods like in the Maya area. If the tropical belt expands climate-wise, which it is expected to do,[12] we can expect concomitant traits to spread as well, including higher temperatures, humidity, and tropical diseases, for instance.

Maya and Archaeology in Belize

The "mysterious Maya." I hear that a lot. The Maya are no more mysterious than any other society. But people seem to think this label fits. It does not. There are unknowns, to be sure, but that's where anthropology comes in—archaeology (past) and sociocultural anthropology (present). Archaeologists do what we can to collect as much information as possible before urban sprawl, industrial agriculture, and looting annihilate Maya history.

Another misunderstanding concerns the Classic Maya "collapse" and the so-called disappearance of the Maya. The Classic period (c. 250–900 CE) political system did indeed fall apart or collapse. Maya kings in the southern lowlands lost power by 900 CE—for good. Kings disappeared; families persevered. And they still do.

Over 7 million Maya live today in Central America and beyond. We also need to rid ourselves of romantic "lost civilization" notions; we are dealing with people who, like others globally, adapted and made mistakes along the way, from which they learned.

The Maya are not timeless. The ancestral and living Maya are not the same. The Maya have always been changing and adapting, like everyone else. The agents of change are diverse—for the ancestral Maya there were the goods, labor, and services they owed to kings as tribute, an expanding population in the seventh through ninth centuries that led to heavier demands on the environment, and climate instability in the form of severe droughts. And after the Spanish invasion into the Maya area in the 1510s, Maya communities faced epidemic diseases the Spanish brought, forced nucleation into villages, forced conversion to Catholicism, and displacement. In the twentieth century, the Maya lived through civil wars throughout Central America, and more recently are living with the land-grabbing of their ancestral lands by non-Maya and a warming climate. Even the term "Maya" obscures the incredible ethnic and local diversity. Today there are 30 known Mayan languages. I use the word "Maya" to make a point and to describe broad practices, for example, traditional agriculture and rituals and their use of forest goods. One final point: "Maya" versus "Mayan." "Mayan" is only used for languages—"Maya" for everything else.

The Maya with whom I work do not feel they are digging up their direct ancestors per se. For instance, while excavating one of the two ancient commoner houses at Saturday Creek in central Belize (600 BCE to 1500s CE), a student asked excavation assistant Juan Antonio if he minded excavating burials. He replied, "No, because they're dead." Juan Antonio's response reflects the dissonance between the past and present because of all that has transpired.

Added to this dissonance with the past is that in Belize Maya do not have legal rights to ancestral sites and places. Until recently, Maya history was not taught much in schools in Belize and elsewhere; the focus had been on colonial history up through the present. That said, the Maya with whom I work are interested in their past largely in the same sense archaeologists and students are.[13]

Nor can we romanticize the Maya. They were not and are not perfect. The archaeological record shows that erosion occurred in the past and that some areas were deforested. However, they adapted. And they fed more people in the past without the widespread deforestation we see today. See for yourself. Look up on Google Earth "Guatemala-Mexico border"; use the historical imagery option and toggle back and forth to see how quickly deforestation is happening. Pollen records from sediment cores extracted from lakes, *cenotes* (steep-sided sinkholes fed by groundwater), and *aguadas* (rainfall-fed natural depressions) throughout the southern lowlands suggest varying degrees of deforestation and forest management or collaboration near the hundreds of Maya cities.[14]

Between globalization, missionization and new technologies, fewer and fewer Maya live the traditional nonanthropocentric, inclusive worldview. Today the Maya often have few options but to adopt "modern" practices to feed their families.

***

Throughout this book I mention sites where I have excavated over the years as director of the Valley of Peace Archaeology (VOPA) project beginning in 1997 (Figure 2).[15] There is the small city of Saturday Creek located along the Belize River, where the Maya lived from c. 600 BCE through the early 1500s CE, that flourished—without kings. There is the medium-sized royal city of Yalbac, where the Maya lived from c. 300 BCE through 900 CE near Yalbac Creek. And then there is Cara Blanca, a pilgrimage destination with 25 pools (lakes and *cenotes*) where the ancestral Maya intensified their ceremonial visits during a period with several prolonged droughts (800–900 CE). I am currently leading a salvage archaeology program to collect as much information as possible from exposed mounds (c. 600 BCE to at least 1100 CE) in recently cleared and plowed areas between Yalbac and Cara Blanca.

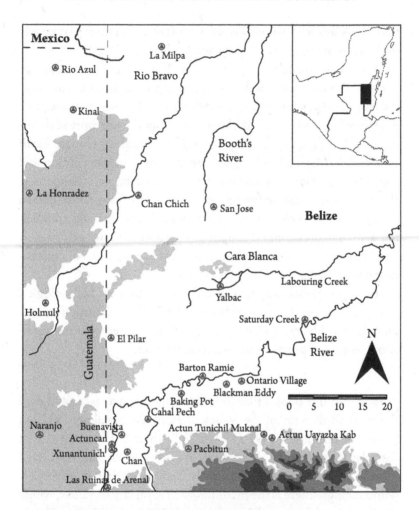

FIGURE 2 *Maya sites in Belize, Central America*
Source: Courtesy of VOPA.

The best part about conducting archaeology in Belize is the people. Maya foremen and excavation assistants are essential for exploring, mapping, and excavating sites (Figure 3). Their stories and insights about living in the tropics are found throughout this book. They help me teach students how to conduct archaeology and about the

FIGURE 3 *2016 crew*

Maya associates include, from the left in the back row, Juan Antonio, Stanley, Cleofo, and peeking out, Javier Arteaga and Javier Gill. Front row left to right: Marcial Arteaga, Carlos Vasquez, Antonio Luna, and Ernesto. I am kneeling in the front backed up by graduate students Aimée Carbaugh, Erin Benson, and Jeannie Larmon and another fab group of field school students.

Source: Courtesy of VOPA.

Maya world—and, if so desired, how to use a machete, the Maya Swiss army knife. They use it for lots of things, from cutting down trees to peeling an orange. And one can practice one's Spanish and experience the beauty of Mayan languages, like Mopan and Q'eqchi'.

In the field, Maya associates do lots more than help excavate and cut swaths through the jungle with machetes. They construct ladders from trees for taking overview photos and for getting in and out of deep excavation pits. They make unit stakes, screen racks, and tables using branches and vines. They use corozo leaf (*Attalea cohune*) and wood posts to make *palapas*, or open-sided thatched structures for shade and to protect us from the elements. Cleofo makes excavation tools out of *sin cheb'* bamboo to excavate burials;

they are quite hard and durable but don't scratch bones like metal tools do.

I have worked the longest with Maya from the Valley of Peace Village in central Belize, located 15 km (9+ miles) north of Belmopan, the capital of Belize.[16] The government of Belize and the United Nations High Commission for Refugees established this village in 1982 as a permanent refugee village during a tumultuous time. Civil wars in Guatemala and El Salvador were particularly harsh on the Maya, and many were massacred. Consequently, this village consists of an amalgamation of nationalities and ethnicities (e.g., Q'eqchi', Mopan, Spanish-speaking Mestizx). To help them assimilate, the Belize government invited 20 Belizean families to live in the village, giving them each a plot of land. Several of these families were Creole—descendants of enslaved Africans. In fact, my first VOPA excavation assistant was Zedekiah Scott, a Creole who grew up in Belize City, Belize's largest city and former capital.

Maya traditional knowledge continues to hold firm and impacts how people engage with the surrounding landscape. Villagers don't need to go to the same church or speak the same language to accomplish this. I have also learned that nearly everything has two to three names—in Spanish, English, Kriol, and/or Mayan—for example, *relleno negro*, *chilmole*, and black *mole* all refer to the same delicious black soup made with chicken, black *recado* (charred tortillas), onion, garlic, salt, tomatoes, and ground pork meat balls enveloping hardboiled chicken eggs.

***

In *Maya Wisdom and the Survival of Our Planet*, I take you on a journey through the tropical jungles of Central America with its plethora of inhabitants and high biodiversity (Chapter 1), followed by a foray into Maya history, highlighting their resilience and their engagement with other Mesoamerican groups, ending with the Spanish invasion (Chapter 2). Chapter 3 introduces Chahk, the Rain God, vital for the rainfall-dependent Maya, and his capricious nature—too much rain, not enough rain—as well as how the Maya adapted to the annual dry season by constructing self-cleaning reservoirs. The stage is now set to introduce the Maya inclusive, nonanthropocentric worldview (Chapter 4), where everything is animated and connected and where the Maya are part of the tropical world, not separate from it. It impacted how they engaged with their environment through reciprocal relations via renewal ceremonies, forest collaboration, pilgrimage to watery portals to the Underworld (Chapter 5), and sustainable agriculture that mimics forest biodiversity (Chapter 6). This engagement begins in the home, the building blocks of society (Chapter 7). Chapter 8, on Maya kings, focuses on their role as water managers who performed the essential rituals to gods and ancestors to provide clean water, a balancing act that worked for over 1,000 years until c. 900 CE. The Maya world also included green cities, some over 2,000 years old, where fluid urban-rural ties were the norm, which are currently being destroyed by looting and encroachment (Chapter 9). Chapter 10 presents concrete solutions crucial for the survival of our planet.

The time to act is now. Everyone can make a difference. Let's begin.

Chapter 1

Setting the Stage

Then and Now

The three realms of the Maya include the 13 sectors of the Upperworld, the Earth's surface, and the 9 sectors of the Underworld. I begin here on the Earth's surface, where the Maya have lived for thousands of years, specifically in the tropical forest, where the "forest belongs to the Maya and they belong to it."[1] I take you on a journey through this realm to show you what living in the tropics entails and what it would have been like for the ancestral Maya.

The tropical world is one of seasonal plentiful rains and drought, high but dispersed biodiversity, hurricanes, periodic prolonged dry periods, and endemic diseases including malaria and carnivorous parasites. Currently, the tropics are heavily impacted by climate change. And as temperatures rise, it is likely that the tropical belt will expand—as will its population. The extensive use of non-Maya non-sustainable practices (e.g., large-scale deforestation, monocropping, and chemical fertilizers, herbicides, and pesticides) also impacts this tropical realm.

The jungle tells its own story through its beauty, aromas, sounds, and wonders. It is home to exquisite orchids and other vibrant flowers; towering hardwoods including cedar (*Cedrela odorata*), logwood (*Haematoxylon campechianum*) from which red dye is extracted, rosewood and mahogany; and a plethora of animals, birds, and reptiles. And insects. Lots of insects. It is odd to envision orchids, considered some of the most beautiful flowers in the world, growing and blooming in dead or dying vegetation (their sustenance comes from decomposing matter rather than soil). Death begets life; we see this often in the Maya world.

You cannot get away from the cacophony of jungle sounds—the myriad of bird calls such as when flocks of parrots roost at dusk, the chirping and buzzing of insects like the deafening cicadas that emerge from their underground nests every 10 years or so, and many other sounds, some probably best left unidentified. And then there are the howler monkeys (*Alouatta pigra*), who obtained their moniker for their resounding and far-reaching calls. You can estimate how far away they are—or how close they are—based on how their howls reverberate. The worst is when they are directly over you. They sometimes throw branches. And feces. Or urinate from the trees. Or just stare at you in absolute silence.

At night, the darkness is complete—on moonless nights, you can't even see your own hand in front of your face. The night sounds are incredible; they're with you when you go to sleep and there when you awake. There is a beauty about the tropical forest, as well as something inscrutable. It is here where the history of the Maya unfolds.

Experiencing the Tropical Forest

Much of the southern lowlands are composed of limestone hills and ridges covered by deciduous hardwoods (mahogany, redwood, logwood, sapodilla, ceiba, purple heart, *chechem* or poisonwood, cedar, pine, allspice, and many more) that are in high demand globally. It comprises two realms—the wet and the dry. The annual five-month dry season turns the jungle into a green desert. The annual seven-month wet season turns it into a swampy and saturated green garden perfect for farming. There is relatively little surface water since much of the rain percolates through the porous limestone bedrock, particularly in deforested areas, and what surface water remains becomes desiccated during the dry season. The tropical world is a force of its own. The washed-out roads and bridges after tropical storms attest to this fact. Another example of this is not being able to get GPS readings because satellites cannot penetrate the forest canopy.

Trekking through the jungle is always an adventure. It is impossible to capture the depths of the tropical forest with a camera. Photos always come out so flat. In many B-movies supposedly set in the jungle, the main characters always seem to run through the "thick tropical vegetation," usually a banana grove. But tropical trees and plants don't naturally grow in clusters or groves but are scattered throughout the forest. It is also at least 10 degrees cooler under the forest canopy than standing in direct sunlight. Secondary growth, that is, postdisturbance growth (e.g., an abandoned field or *milpa*), is much worse; it is thick, hot, humid, heavy, and difficult to work in. Conducting archaeology in the hot and humid tropics is both challenging and rewarding. It is easy for the inexperienced to get lost. Compasses work, but unless you track orientation and distance, it is moot when surrounded by greenery that looks the same on all sides.

That is just how the jungle is—impossible to understand until you've been standing in the humidity, dripping sweat, with red ants crawling up your legs. Long sleeves and hats are a must. There are too many things that cut, tear, and bite. Thick cloth is the best. Thin cloth might be cooler, but you will soon regret this decision as you begin to feel mosquitoes biting.

The jungle floor is covered with debris—leaves, vines, fallen trees and other vegetation, lots of seemingly invisible bugs (chiggers), and all that is left of completely decomposed trees—holes in the ground hidden by debris that you usually only realize are there when you have stepped into one. But if you lose your footing, don't grab any trees! Some have spikes, some have thorns, some have resin that causes rashes or worse, some are home to termites, ants, spiders, snakes, and other creatures. That said, there are just as many botanical medicines that treat bites, rashes, fevers, illnesses, and other ailments.

My point is that everything is so diverse—and connected. And this diversity and connectivity played a major role in ancestral Maya life and continue to play a role globally. In fact, tropical forests play a vital role on Earth. Trees, for example, connect the three major realms—the Upperworld, Earth, and the Underworld and "constitute the environmental quality committee—running air and water purification service 24/7."[2] Trees produce oxygen. Trees absorb carbon dioxide (CO_2). Trees support life. "A single tree in the Peruvian Amazon may be home to more ant species than the entire British Isles...and more tree species can be found in less than one square kilometer of tropical rainforest in Malaysia than in all of the United States and Canada."[3] Forests act as windbreakers, limiting the impacts of forceful winds, as well as floods.[4] The erosion of soils due to wind hastens fertility loss.

A treeless landscape decreases biodiversity, increases soil erosion, and reduces photosynthesis and CO_2 absorption. Deforestation exacerbates the spread of pests and results in increased temperatures and humidity and decreased amounts of rainfall since deforestation interferes with evapotranspiration (i.e., the evaporation of water from soils and plants into the atmosphere) because water is released into the limestone bedrock rather than into the atmosphere via leaves and roots.[5] Indeed, a recent study using satellite data of forest loss and precipitation in the Congo Basin, the Amazon Basin, and Southeast Asia between 2003 and 2017 shows a pattern of reduced rainfall, up to 0.25 ± 0.1 mm (less than 1/10 of an inch) per month for every percentage point reduction in forest cover.[6] Deforestation increases aquatic sedimentation and decreases water quality. Droughts, which are becoming more extreme due to global climate change, increase the susceptibility of trees to disease, insects, and wildfires.

Everything blossoms at different times in the jungle. So fresh "produce" is always available. Traditional agricultural practices mirror this pulsating system. The flora and fauna illustrate not only the amazing biodiversity, but also what kinds of resources the Maya had at their disposal, which they clearly made use of—sustainably. We know this because animal bones representing diverse species never disappear from the archaeological record, even in the Late Classic period (c. 600–900 CE), when Maya populations were at their highest. What does this signify? That even when there were hundreds of Maya cities and tens of millions of farmers, the forest never disappeared. We know this because wild fauna require the forest to breed and thrive. Contrary to what some think, the Maya did not denude their landscape of vegetation. They *coexisted*, and quite well. A reminder: no plants or animals became extinct during this type of coexistence.

Jungle Life

Biodiversity in the jungle is astounding. A plethora of flora are depicted in Classic Maya iconography—on ceramics, monuments, portable items, and textiles, including copal (*Protium copal*), the resin of which is burned as incense in every ceremony, sapodilla trees (*Manilkara zapota*) for their edible fruit and resin (*chicle* or gum), water lilies (*Nymphaea ampla*), which indicate clean water, allspice (*Pimenta dioica*) for its aromatic seeds and leaves, and more. Even the sarcophagus of Palenque's most powerful king, K'inich Janaab' Pakal (603–683 CE), was decorated with avocado (*Persea americana* Mill.) trees and the tree of life, the ceiba (*Ceiba* sp. Mill.), and other life-giving vegetation.

Tropical vines leave their own meandering stamp. There are the strangling vine and the water vine. Strangling vines are a sight to behold, albeit sometimes an unsettling one (e.g., *Ficus aurea* and *Ficus cotinifolia*). They literally squeeze the life out of the tree as they grow and ultimately replace the tree. The water vine, or water tie-tie in Kriol (*Vitis tiliifolia*), instead of resulting in death results in life. The water vine consists of a vine over 5 cm in diameter (2+ inches) sloping horizontally between trees. Once you find one, you slice it through with a machete. You hold up the vine, and out comes filtered water that is some of the most refreshing water you will ever drink. Or maybe it is because you are parched, and anything tastes good. The strangling and water vines are quite different, yet both serve essential roles in maintaining biodiversity.

Rather than just list hundreds and hundreds of plants and animals, I illustrate this biodiversity using the Choc family home garden as a microcosm (Figure 1.1). As the saying goes, a picture is worth a thousand words.

The Choc home garden includes a mix of wild plants and native and nonnative domesticates—and chickens and pigs wandering about.[7] At least 65 different species of plants, trees, and flowers are shown on the map. They encompass a broad range of uses—for consumption (e.g., fruits, vegetables, palm hearts, nuts, spices, garnishes, and herbs), medicinal uses (for pain, "to improve the blood"), construction (timber and vines for rope), household use (insect repellent, baskets, and leaves to wrap food), ornaments (flowers), and poison (to stun fish). Their *milpa* (field) is less diverse these days: maize, beans, cassava, and okra. Cleofo used to plant maize, beans, chaya, papaya, cassava, potatoes, coco, and sometimes rice and pineapple. Such gardens and fields are a major reason the Maya have such a long history, because they mimic a diverse forest.

To explore how the ancestral Maya collaborated and learned from the forest, Cleofo and former graduate student Colleen Lindsay collected over 300 botanical species from areas with and without ancient Maya sites in the Valley of Peace Archaeology (VOPA) project area in central Belize that had been unoccupied for over 1,000 years. The goal was to assess if the forest today reflects ancient forest collaboration rather than a primary forest per se. It does. They found evidence that the Maya promoted some botanical species over others, especially near residences and cities—those that had medicinal properties, were edible, and could be used in construction, or ones that attracted particular kinds of animals or birds.[8]

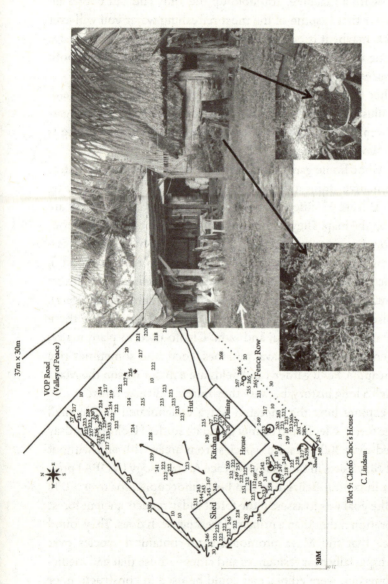

FIGURE 1.1 Choc family home and garden

The numbers on the map signify different species of flora. *Maxik* (bird chili peppers, *Solanum* sp. *Solanaceae*) are in the photo on the left, and *yerbabrena* (Spanish) or *iske* (Q'eqchi') or mint on the right. The hollowed-out tree stump in front of the kitchen is used to churn butter.

Source: Courtesy of VOPA.

Cleofo was able to identify 95% of the species in English, Spanish, Mopan, or Q'eqchi'. They all had uses—as fuel, botanical medicines, construction materials, edible plants, nuts and fruits, and more. The five percent he could not identify "had no use"; thus, they had no name other than "plant" or "tree" (the same goes for reptiles, animals, birds, fish, turtles, crustaceans, and mollusks). The forest we see today is literally a Maya forest because they collaborated *with* it in a sustainable manner. The same goes for animals, which the Maya also depicted in their iconography—howler monkeys, jaguars, spider monkeys, foxes, bats, deer, peccary, tapir, iguanas, armadillos, opossums, anteaters, shrews, the striped basilisk lizard, coatimundi, crocodiles, rabbits, pumas, jaguarundis, margays, ocelots, skunks, red brockets, dogs, Gulf Coast toads, Mexican tree frogs, lots of rodents, and snakes.

The king of the jungle is the jaguar (*Panthera onca*). Maya kings wore the pelts of these powerful cats or draped them on thrones. They signify grace and power. The jaguar is associated with the sun, especially as it travels through the Underworld every night before it once again emerges at dawn. Birds feature prominently in Maya iconography, tales, and stories (e.g., hummingbirds, falcons, ornate hawk-eagles, woodpeckers, rainbow-billed toucans, vultures, blue-crowned motmots, collared aracari toucans, parrots, slaty-tailed trogons, oscillated turkeys, resplendent quetzal birds, and scarlet macaws). Iridescent green feathers of the quetzal bird (*Pharomachrus m. mocinno*) from highland Guatemala are used in royal headdresses. The scarlet macaw (*Ara macao*) plays a role in the Maya origin history as the false sun. The vermiculated screech-owl (*Otus guatemalae*) is a portent of death and a messenger from the Underworld. All are losing their homes due to land clearing for agriculture, ranches, and settlements.

Freshwater and marine life are integral to Maya life too, because they are part of the watery Underworld. Archaeologists find coral, queen conch and thorny oyster shells, stingray spines, shark teeth, and marine fossils in the archaeological record and in the iconography. We also find fish bones in middens (a fancy term for ancient trash).

Like everything else in the tropical forest, the Maya engaged with animals in multiple ways. For example, in addition to the slider turtle (*Trachemys* [*Pseudemys*] *scripta*) providing an important source of meat and calcium (eggs), it is depicted in the iconography, with the reborn Maize God emerging from its carapace, which signifies the surface of the Earth.[9] Turtle carapaces also are used in ceremonies (drilled turtle shells) and medicinally—ground turtle shell mixed with water is given to babies to ease asthma and coughing. Nearly every part of the turtle is used. What is not used nourishes forests, fields, or gardens. Nothing is wasted.

Diseases, Snakes, and Bugs, Oh My

It's the little things that matter.

The ancestral Maya dealt with numerous small creatures that continue to leave their mark. Mosquitoes, scorpions, snakes, and others are depicted in their iconography. There are lots of insects that bite—doctor flies, or *tabanos*, all kinds of flying insects—including flies, mosquitoes, bees, wasps, and more. There are scorpions, snakes of all kinds (e.g., coral snake, fer-de-lance, and boa constrictor), termite nests (often taken over by black ants), lion trap spiders, tarantulas, large flying cockroaches, other cockroaches of all sizes, worms, and millipedes. Insects play a larger role than you might think. In the tropics, termites (*Isoptera*) are analogous to earthworms in temperate areas—that is, they are beneficial. That said, by-products of termite cellulose digestion include CO_2, methane, and hydrogen gases that are released into the atmosphere, a process made worse in wet savannas and disturbed areas such as cleared and burned fields.[10]

While the ancestral Maya did not have to contend with epidemic diseases until the Spanish and other Europeans brought them beginning in the late fifteenth century, they did have to deal with endemic diseases, such as hepatic schistosomiasis, a liver disease caused by a parasitic worm passed to humans via infected snails in contaminated water.[11] There are other disease carriers, like the dangerous kissing bug (*Triatoma* sp.), which carries the parasite that causes Chagas disease, which typically results in a fever and, if left untreated, can lead to heart disease. Today, stagnant water provides ideal conditions for parasites and water-borne diseases to bloom, such as cholera and other diseases that cause diarrhea and other detrimental symptoms. In fact, the more heavily settled an area is, the more insects and other diseases spread and multiply. Temperature variation throughout the year is not extreme enough to kill off pests.

Parasites are one reason tropical areas in the Americas lacked large native livestock and beasts of burden. When the Europeans introduced cattle, horses, and other livestock, they lost many to parasites. It wasn't until about 20 years ago that Belizeans were able to prevent most of their livestock from getting screwworms (*Cochliomyia hominivorax*), flesh-eating larvae whose eggs are deposited by flies in open wounds.

Mosquitoes are the bane of the tropics. Sometimes there are none about, like at the end of the dry season when it hasn't rained for months. Then there are thick clouds of them soon after the rains begin. They don't like direct sun or too much breeze or smoke. They do, however, like shaded areas without too much breeze. Repellent with 25% Deet works the best at keeping them at bay. Cleofo swears that eating chili peppers every day repels mosquitoes.

And then there are tick bombs. Wherever livestock is, ticks are nearby. Hundreds of tiny ticks the size of freckles await on the underside of leaves for warm-blooded creatures to walk by. And we cannot forget the bot fly (*Dermatobia hominis*). Everyone has their own ideas for how to get rid of bot fly larvae, locally called *comayoté*. Eggs are deposited when you are bitten by parasite-ridden mosquitoes. As larvae grow underneath your skin, they eat flesh—your flesh. It feels like someone is pinching you hard. Because they require a breathing hole, the common way to kill them is to suffocate them by putting toothpaste, packing tape, or tobacco over the hole. Once they are dead, you have to pop them out so their decomposing remains don't infect the same flesh upon which they had previously feasted.

But there are lots of beautiful butterflies of all kinds (e.g., blue *Morpho peleides*). The tropical forest abounds with beautiful life—some quite dangerous. Brightly colored insects and reptiles that stand out are poisonous or deadly—frogs, millipedes, and snakes. A great example is the coral snake (*Micrurus diastema*). Coral snakes are beautiful—red, black, and yellow—and they are neurotoxic. There is no antidote—but they have tiny mouths unlikely to bite anything larger than a bee.

While books such as *A Field Guide to the Snakes of Belize*[12] are useful to determine venomous (hemotoxic and neurotoxic) versus harmless snakes, it is best to rely on the expertise of the Maya. Most common is the hemotoxic fer-de-lance (*Bothrops atrox asper*), also known as yellow jaw, tommygoff, or *barba amarilla*. Snake-bite kits and machetes are a must since antivenins require refrigeration, something not really feasible while exploring or excavating in the jungle. Interestingly, snakes are depicted in Maya iconography more than any other creature—to record vision quests, for example. There are nonfatal and fatal interactions between snakes and people, especially involving the fer-de-lance. The potency of its venom depends on various things. Young fer-de-lance snakes have not yet learned to control how much venom they expend, so they can be particularly dangerous. Pregnant females are also quite dangerous. Since they digest and gestate in the same place, their venom builds up. And they are more protective and doubly dangerous. Fer-de-lance snakes are the least dangerous right after a kill when they have expended their venom.

There are lots of different kinds of spiders. Tarantulas are some of the largest. Scorpions of different sizes are like chili peppers—the smaller they are, the more bite they have. But larger ones look scarier (e.g., *Centruroides gracilis*). There are innumerable species of ants. They all bite. Subin tree (*Acacia hindsii*) branches are covered in large hollow thorns in which ants (*Pseudomyrmex ferrugineus*) live. When anything touches any part of the tree, the ants come out in droves and immediately attack whatever creature has dared touch or land on their home. And they release formic acid when they bite.

Leafcutter ants (*Atta cephalotes*) are very industrious—and relaxing to watch. Their 8-cm (3+ inches)-wide trails through the forest floor are always so clean—no debris whatsoever. They bring their freshly cut leaf bits to their underground home as sustenance for the fungi that their larvae eat. They help regenerate soils because the leaves break down and create compost.

As for army ants (probably *Eciton burchelli*; there are many species), the best thing you can do is to just get out of their way; they have a nasty bite. When they leave their nest, they do so in the tens of thousands, if not more, creating a 1- to 2-m-wide path in the jungle—a living floor. They attack anything up to the size of a kitten. The only creature they avoid is a florescent orange centipede that comes out in droves at the beginning of the rainy season. Why? Because it is poisonous. The few times our excavations have been invaded by army ants, they soon became bored with our pristine excavation units and departed.

The ancestral Maya kept stingless bees (*Melipona beecheii*) to make *balché*, a fermented drink made from honey and bark from the *ba'al che'* tree (*Lonchocarpus yucatanensis*), and used beehives for wax. The Maya continue to keep honeybees and collect wild honey. Bees also help pollinate maize, beans, squash, and other crops. Killer, or "Africanized," bees—a subspecies of *Apis mellifera*—showed up in Belize in the late 1980s (and in Texas in the 1990s and Florida in 2002). This was an unintended consequence of human action. African bees were brought over to breed with South American honeybees in the 1950s because they produce more honey. But some escaped. African queen bees took over hives, propagated, and spread.

There's a multitude of other ant species, termites, spiders, cockroaches, biting insects, gnats, flies, hornets, bees, wasps, centipedes, millipedes, ticks, chiggers, grasshoppers, slugs, worms, butterflies, and lots more that are found on/over/under nearly every jungle leaf on branches or those littering the ground. An important lesson is to never put food on any surface. Ants and other insects will devour it; hang it in a bag from a tree limb. And don't sit on the bare ground either. There are too many things we can't see, but they sure as heck bite.

It's the little things that matter. Insects pollinate plants, aerate soils, control pests (e.g., dragonflies consume mosquitoes and their larvae), and scavenge the dead recycling nutrients into the soil in the process. Their diversity is astounding and directly mimics the diversity of other jungle life.

Biodiversity is decreasing though, and fast; this is not a good thing since biodiversity provisions the very ecosystems on which we rely—water, food, and air.[13]

Non-Maya Practices

Several U.S. agricultural specialists have come to Belize over the years to test novel agricultural techniques that often do not work, likely because the specialists are unfamiliar with tropical conditions, soil types, and weather. The real reason their novel ideas do not reach fruition is because they never talk to local Maya farmers. This is why their ideas fail more often than not. Spanish Lookout modernized Mennonite farmers fall into this category. Despite their appreciation of Maya knowledge ("Da Maya, dey smarter dan we"), farmers clear-cut and plow the jungle and mounds alike, erasing forests and Maya history. While massive land clearing and monocropping produce high agricultural yields, they are unsustainable. The Maya have been farming for 4,000 years without using monocropping or causing massive deforestation and fed more people in the past that current non-Maya farming practices do at present.

Spanish Lookout Mennonites, unlike the three other Mennonite communities that first arrived in Belize in the 1950s, began modernizing in 1961. No more horse and buggy. No more men with beards without mustaches, wearing straw hats and suspenders. No more women wearing long-sleeved, long black dresses with black bonnets. Spanish Lookout farmers invested full-on in heavy machinery, electricity, phones, and more—including large-scale monocrop agriculture and the use of chemical pesticides, herbicides, and fertilizers.

There is also a long history of logging in Belize—for Honduran mahogany (*Swietenia macrophylla*), for example, and many other hardwoods (e.g., cedar, redwood, and purple heart).[14] Yalbac Ranch, which until recently owned the property where Yalbac and Cara Blanca are located, had been sustainably logging for decades; you didn't see clear-cutting anywhere. And they planted four trees for every tree they cut. This all changed after the October 2010 Hurricane Richard when central Belize was ground zero. The destruction to the hardwoods was heartbreaking and devasting for sustainable logging.

In February 2011, while in Belize assessing the hurricane's damage to Yalbac and Cara Blanca at the behest of the Belize Institute of Archaeology, I stood on the Lookout, a cliff overlooking the western Cara Blanca pools. It looked as if the gods had used a giant lawnmower to mow the jungle. Yalbac Ranch had no choice but to sell over 121 km² (47 square miles), which they sold to the Spanish Lookout Community Corporation. This led to the clear-cutting of a thousand or so hectares (2,471 acres) in 2014, exposing hundreds of ancient Maya mounds (Figure 1.2). They continue to clearcut thousands of acres every year. They typically use two large bulldozers with a massive chain linked between them to shear off anything green. Once all the dead vegetation dries out and after they remove logs to sell, they burn the rest. Animals that are not fast enough to escape, like turtles, are burned alive. And while the farmers are supposed to plow around the taller mounds, they often do not.

As mentioned, the Mennonites only buy or lease land with lots of Maya mounds because they know the Maya lived on fertile agricultural soils. So, each time they plow, they erase one to two generations of history if not more. Generations of Maya families lived in the same place, renewing everything every 20 years or so. Mounds thus are like mini-tells with a multitude of layers representing hundreds of years of family histories and information about sustainable

FIGURE 1.2 *Top:* looking southwest from the clifftop immediately above Cara Blanca Pool 7 (see Figure 5.1 for a map of the pools) before 2010 Hurricane Richard. *Bottom:* a few years after the hurricane, taken from the same spot.

Source: Photos by the author.

practices. The Mennonites also carve up hills for roads, which means easier access for poaching, illegal logging, and looting. The building of new roads always indicates that massive clearing and large-scale and extensive farming will soon follow.

In 2016 and again in 2022, with funding from the National Science Foundation and with permission from the community leaders of Spanish Lookout and the Belize Institute of Archaeology, we conducted a salvage archaeology project in the recently cleared areas to collect as much information as possible before the ancient Maya houses are gone.[15] Most artifacts near the surface date to the Late Classic (c. 600–800 CE) or Terminal Classic (c. 800–900 CE), but we do not know how much history has been lost to plowing.[16] At a small mound near one of the Cara Blanca lakes in 2016, for instance, we found two early Postclassic (c. 900–1100 CE or later) arrow points, indicating that the Maya still lived in this area, near lakes anyway, after they had abandoned kings and cities by 900 CE due to a series of events set in motion by several prolonged droughts (more about this in Chapter 2).

***

Welcome to the jungle.

You should now have a sense of what the jungle was like for the ancestral Maya, as well as its current condition and potential future. This was the world the ancestors of the Maya experienced when they first arrived about 13,000 years ago.

Chapter 2

The Maya

May the water be taken away, emptied out, so that the plate of the earth may be created—may it be gathered and become level. Then may it be sown; then may the dawn the sky and the earth. There can be no worship, no reverence given by what we have framed and what we have shaped, until humanity has been created, until people have been made

POPOL VUH, *sixteenth century K'iche' Maya origin history*[1]

A savanna. That's what central Belize was like 27,000 years ago during the Late Pleistocene and the Last Glacial Maximum—and before the arrival of humans. Water was scarce due to longer dry seasons and less rainfall. Cara Blanca's steep-sided sinkholes, or *cenotes*, had some water—well, at least the deeper ones did. Groundwater levels were low due to lower sea levels; much of the seawater was taken up as glacial ice. Savanna inhabitants, such as giant sloths (*Eremotherium laurillardi*), had to climb down at least 21 m (69 ft) into the steep-sided sinkhole (Pool 1) to reach water to quench their thirst. The thing is, they couldn't get back out; the sinkhole was just too steep and deep. Trapped, they eventually starved to death. These sloths were giant indeed; they could grow up to over 6 m (20 ft) in length and weigh over 7 tons. And they weren't alone. Fossil-laden clay beds encircle the entirety of Pool 1—who knows how many other megafauna species became entombed.

Pool 1 (100 × 70 m, or 328 × 230 ft) water levels are much higher (62+ m or 200+ ft, deep) today and thus much closer to the ground surface. In 2014, with funding from the National Science Foundation, paleontologist M. Gregory McDonald and cave exploration diver Chip Petersen extracted a giant sloth tooth from a clay bed 21 m deep. Using carbon and oxygen isotope analysis of the tooth, Larmon and colleagues were able to reconstruct not only how the animal's diet shifted from the wet to the dry season, but the environment itself.[2] This tropical savanna had a long dry season of about nine months that stands in stark contrast to the current tropical forest with a five-month dry season.

It was into this world that the First Americans arrived about 20,000 years ago in the Americas, long before Christopher Columbus. They saw a far different world than today. They saw megafauna from the size of elephants to smaller species the size of a cow, like the armored relative of the armadillo, the glyptodont. Along with the familiar mammoth, they saw a strange-looking relative, the gomphothere with its long, straight tusks. They also encountered a variety of large predators such as saber-tooth cats and the huge short-faced bear. It is amazing that these giants roamed the Americas from 2 million years ago until about 10,000 years ago. That's when they became extinct at the end of the last Ice Age. Why? The extinction of megafauna likely resulted from a drastically changing climate and new immigrants better adapted to the new world—humans.[3]

This long history in the Americas is corroborated by Western science and Indigenous origin histories. Robin Wall Kimmerer, an enrolled member of the Citizen Potowatomi Nation and a SUNY distinguished teaching professor and director of the Center for Native Peoples and the Environment, tells of Skywoman, who fell from Skyworld. She is the Mother of the People and "was the first immigrant." Nanabozho, the Anishinaabe culture hero, First

Man and Original Man, embodies the fact that "we humans are the newest arrivals on earth, the youngsters, just learning to find our way."[4]

Flora and fauna have always interacted with each other, shaped each other. Then humans were added to the mix. They have coexisted, influenced, and coevolved with innumerable flora and fauna species. Humans, however, including the Maya, always leave a more palpable footprint.

In what follows, I focus on brief moments in time, whether they concern climate, human history, or population size, to highlight clues to how we can divert from our current anthropocentric course to one that ensures the survival of our planet.

Mesoamerica

It is easy to think about the Maya in isolation. However, they did not emerge or exist in a vacuum but were part of a complex and interwoven history with other Mesoamerican peoples and cultures. Mesoamerica encompasses diverse environmental zones and ethnicities and extends from the Southwestern United States through Honduras in Central America. This complex relationship began about 20,000 years ago when the First Americans started arriving via Beringia, the land bridge that existed between Russia and Alaska during the last Ice Age, which is now the Bering Strait, or later along the coast in small boats.

Mesoamerica also embodies shared elements that include an inclusive worldview and a world where hills and mountains were homes to ancestors and caves and water bodies were portals to the Underworld. Jade, signifying water, was their gold. Animal companion spirits connected ancestors, forests, and people. Pyramid temples signified water and/or lineage mountains where ancestors resided. Staple foods consisted of maize, beans, and squash. Their origin histories had common themes including the role of the Hero Twins and a universe with three major planes (Upperworld, Earth, and Underworld). Most cities had ball courts with multiple purposes—for ball games between cities, friends, and foes; to re-enact origin histories; and as entry points to major urban temple and palace complexes. They also had a worldview that embodied complementary opposition rather than Cartesian dichotomies (see Chapter 4). The essential role that ancestors played in daily life was another shared feature, as were the ceremonies performed by commoners and royals alike to reach ancestors, rain and sun gods and other deities, and forest spirits. And they shared a complex yet intersecting calendrical system encompassing a 260-day ritual calendar and a 365-day solar one. Mesoamerican peoples also were skilled in astronomy—not for the sake of science per se, but for astrological purposes. In other words, Mesoamerican peoples are connected through space and time, including the Maya.

In what follows, I focus on the southern lowland region, which today consists of Belize, northern Guatemala, southeastern Mexico, and western Honduras. It noticeably differs from the northern lowlands encompassing the Mexican states of Yucatán and most of Campeche and Quintana Roo in several essential ways that have resulted in unique but related histories that scholars too often conflate. In general, the southern lowlands have higher elevations (e.g., Tikal, Calakmul, and Caracol range from 245 to 500 m above sea level) and rainfall amounts, more and better agricultural soils, and a deeper and less accessible water table (Table 2.1). While cities in the southern lowlands were abandoned by 900 CE, those in the northern lowlands saw a florescence due in part to an influx of immigrants from the south (e.g., Chichén Itzá) that lasted centuries and continued even during the arrival of the Spanish beginning in 1511.[5] Conflating evidence from these two regions supports a broad range of ideas or hypotheses, which would be less supported when only including evidence from one or the other region.

TABLE 2.1 Major features distinguishing southern and northern Maya lowlands*

	Southern lowlands	Northern lowlands
Elevation (m above sea level)	100–500	North/northeast: <40 South/southwest: 30–200
Annual rainfall (mm)	1,350–3,700	North/northeast: 1,200–1,500 South/southwest: 1,000–1,200
Water table depth (m)	100+	North/northeast: 5–20 South/southwest: 20+
Surface water	*Bajos*, some rivers, lakes, *cenotes* rare†	North/northeast: 7,000+ *cenotes*, freshwater wetlands, lakes South/southwest: some rivers, no *cenotes*
Agricultural soils	Fertile mollisols	North/northeast: thin soils, lots of *rejolladas*‡ South/southwest: well-drained and poorly drained clayey soils
Occupation history	Majority through c. 900	1100s through Spanish Conquest in early 1500s

* Adapted from L. J. Lucero. 2023. Ancient Maya Reservoirs, Constructed Wetlands, and Future Water Needs. *Proceedings of the National Academy of Sciences* 120:e2306870120. doi: 10.1073/pnas.2306870120, Table 1.

† *Cenotes* are found in central Belize because they are only about 64 km (40 miles) from the coast in lower elevations than the interior (e.g., Pool 1 is at 59 m above sea level, or 194 ft).

‡ Karstic sinkholes with high humidity and filled with moist soils excellently suited to cultivate plants and trees.

When Did the Maya Become "Maya"?

This question is challenging to answer since ethnicities and languages leave no visible trace in the archaeological record. That said, linguistic anthropologists estimate that Proto-Mayan language developed around Uspantán in the northern highlands of Guatemala and began diversifying about 4,200 years ago. The story of people in this area began at least 13,000 years ago when ancestors of the Maya first arrived in this tropical world. It is fascinating that megafauna and people overlapped for a brief moment in history—for about 3,000 years or so (Figure 2.1).

For thousands of years, people lived a nomadic existence hunting game and gathering nuts, berries, roots, and other plant foodstuffs. We don't yet see a distinctive Maya fingerprint—for example, you cannot distinguish stone tools found in the Maya area from those found elsewhere in Mesoamerica. This mobile lifestyle left a minimal trace in the archaeological record. Also, many temporary seasonal preagricultural settlements lay deep underneath ancient Maya cities and modern settlements.

Then came the domestication of maize about 8,700 years ago in central Mexico, reaching the Maya area by about 5,500 years ago. Other crops soon followed—for example, manioc (*Manihot esculenta* Crantz) and various chili peppers.[6] People adopted farming very gradually because they had plentiful and diverse forest foodstuffs. Also, farming is hard work. People only adopt agriculture when they need to feed more people in any given area. In about 2000 BCE, the Maya began living in farming communities and began to clear land. Just prior to 1000 BCE, the Maya began producing and using pottery and other items they did not need as nomadic hunters and gatherers. They began to noticeably transform the landscape beginning c. 200 BCE with terraces, canals, and raised fields (e.g., northern Belize). Interestingly, beans did not appear in the Maya area until sometime between c. 300–100 BCE; perhaps the growing population needed more protein as more land was cleared, and they

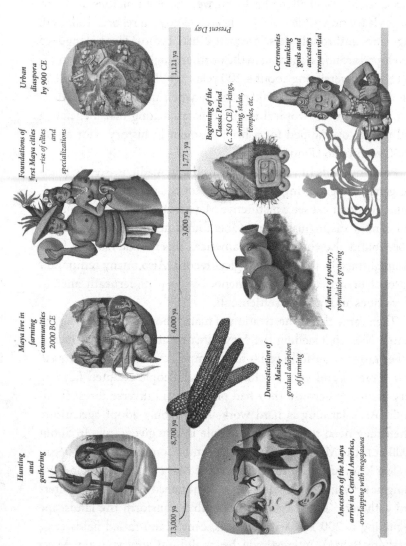

FIGURE 2.1 *Maya history*

Source: Illustration by Reilly Jensen. Courtesy of VOPA.

wanted to maintain the number of available game? The Maya adapted quite well, and the population grew; the archaeological record shows an increasing presence of Maya people throughout the landscape (Figure 2.2).

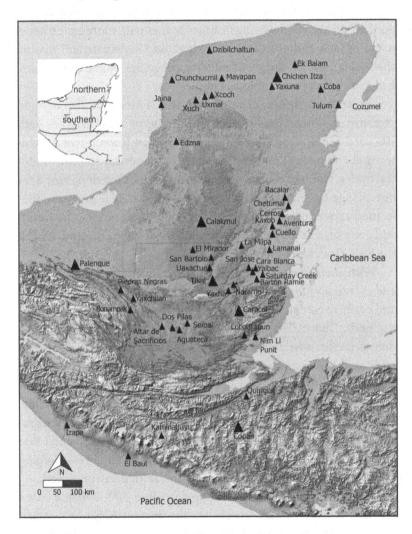

FIGURE 2.2 *Maya cities mentioned in the text*

Source: Courtesy of VOPA. For a map with all recorded Maya sites, see http://www.mayamap.org/.

After 1000 BCE we see inklings of the first cities; communities grew into towns, and towns grew into cities. The Maya also started to branch out as the population continued to grow, moving away from coasts, rivers, and lakes into interior areas such as the Petén in Guatemala. The first settlers became the first elites, who distinguished themselves by living in larger homes with more exotic items made of jade and obsidian from highland Guatemala and marine shell. Elite patrons sponsored building programs, public ceremonies, ball games, and feasts for local families.

By 300 BCE, the Maya "traits" with which we are familiar became established—increasingly larger and ornate pyramid temples with stuccoed and sculpted deity masks, E-Groups (monumental architectural complexes) that commemorate calendrical and astronomical cycles, ball courts, large plazas, and palaces. One thing that does not change throughout their history is the fact that everyone performed ceremonies in the home, focusing on ancestors, the forest, and gods such as the Maize God, K'inich Ajaw the Sun God, Chahk the Rain God, and others.

With increasing population and administrative needs, leaders and eventually kings emerged to allocate resources, resolve disputes, and perform ceremonies that highlighted their closer connections to the gods. The earliest leaders, and later kings, performed public domestic rituals writ large to attract subjects and exact tribute. They replicated and expanded but did not restrict the family-level rituals. These integrative and increasingly larger public ceremonies promoted solidarity, a sense of belonging, and social membership. Everyone had ceremonial duties—from farmers to kings, an integrative tool to promote *popol*, or community. Kings also ensured that cities became part of ceremonial or pilgrimage circuits—merging urban and rural, humans and nonhumans, settled and unsettled.

The earliest known Maya writing dates to c. 300–200 BCE based on the exquisite murals revealed at the ancient Maya city of San Bartolo to the northeast of Tikal in Guatemala.[7] There is even an early glyph of *ajaw* (ruler) dating to c. 100 BCE. The stage was set for the Classic period, when royal power and population attained their greatest extent. Recent lidar mapping has revealed a landscape dotted with cities of all sizes interspersed with farmsteads and fields, as well as areas largely devoid of settlement, signifying forests, *bajos* (seasonal swamps or wetlands), and areas purposefully left untouched (more about this in Chapter 5).

The Classic Maya (c. 250–900 CE)

Unlike ancient cities in temperate zones such as China, Europe, Mesopotamia, Andean South America, and Egypt that arose along major rivers and fertile floodplains, the most powerful Maya cities are not found along rivers or lakes. Instead, the Maya built such cities in areas with some of the richest tropical soils in the world (mollisols), which are found throughout the landscape in variously sized pockets that are up to 100 cm deep.[8] The wet and dry seasons also intertwined with royal histories—and power—because everything depended on the annual rains between mid-June and mid-January. Rains replenished urban reservoirs, nourished forests and crops, and hydrated everyone and everything.

There were hundreds of Maya cities with kings, surrounded by rural farmsteads (solitary residences, or *plazuelas* with two to five structures surrounding an open area), forests, and fertile soils and extensive *bajos* that together supported tens of millions of people. Some kings were more powerful than others, especially those at Tikal and Naranjo Sa'aal in Guatemala, Calakmul in Mexico, and Caracol in Belize. None of these cities are found near lakes or rivers. Early administrators and kings compensated for the lack of surface water by constructing what eventually came to be massive reservoirs (detailed in Chapter 3). Written records, expressed in exquisite hieroglyphs, regale the lives of kings and royal families—their birth, accession, battles, and major ceremonies. And not much else. The study of the ancestral Maya thus is not found in the realm of history departments, but in anthropology or archaeology departments. Cities, though, tell their own stories through their palaces, temples, ball courts, carved monuments, *sak b'eh* (elevated causeways), open plazas, and artificial reservoirs that you can still see today throughout the Maya area.

Rural farmers depended on urban reservoirs during the annual dry season when surface water became scare. They also came for markets, royal ceremonies, ball games, and other large-scale public events. Royals and urban elites depended on rural farmers for labor to build and maintain reservoirs and monumental constructions, services (e.g., craft specialists or hunters), agricultural products (e.g., maize, beans, manioc, squash, chaya, pineapple, tobacco, and tomatoes), and forest products (e.g., wood, fuel, construction materials, medicinal plants, chert, game, and fruit). Royal power thus was somewhat seasonal.

This symphony worked for over a thousand years. Classic Maya kings were excellent water managers—as long as the rains came. Water became increasingly vital, especially in the Late Classic period (c. 600–800 CE) when population numbers and land use peaked. When several prolonged droughts struck the Maya area beginning c. 800 CE and recurring for over a century, reservoir levels plummeted, crops failed, and famine ensued. In most cases, subjects did not revolt or resort to violence. They voted with their feet. Maya farmers-slash-subjects deserted kings and southern lowland cities by 900 CE to find more stable water supplies. They left their ancestral homes for the sake of their families' survival.

Classic Maya kings disappeared. Farmers adapted and moved on; they always do.

What Happened after the Fall of Maya Kings?

An urban diaspora happened. Maya families left the interior, going in all directions, settling in smaller communities throughout the

Maya area, as well as in cities beyond the southern lowlands. While this response was drastic, it was an adaptive strategy and one that worked, as evidenced by the over 7 million Maya currently living in Central America and beyond.

About 90% of farmers left the southern lowland interior between 800 and 900 CE for coastal areas, highland Guatemala, the northern lowlands, and areas near major rivers or larger lakes (e.g., Belize River, Lake Petén Itzá in Guatemala).[9] For example, in the northern lowlands, maritime trade grew along with coastal cities and accessible interior cities, including Chichén Iztá and later Mayapan, Chetumal, Cozumel, Bacalar, and Tulum. Market towns and trade thrived. The families who remained in the southern lowland interior lived near the relatively few perennial lakes and rivers (e.g., Lake Petén Iztá, Lake Yaxha, Belize River, Cara Blanca lakes) with a different, smaller sociopolitical organization. Some even remained in the southern interior cities, relying on the unmaintained reservoirs for centuries.

The northern lowlands witnessed a florescence due to the influx of immigrants from the south in the Terminal Classic (c. 800–900 CE) and Early Postclassic (c. 900–1200 CE) periods. There, the newcomers thrived under a new and different political system—shared rulership rather than sole rulership, as had been the case for Late Classic kings in the southern lowlands. The feathered serpent K'uk'ulcan (or Quetzalcoatl, the Nahuatl version from central Mexico) symbolized this shared power, as did other symbols from the Nahuatl-speaking Toltecs from the city of Tula (c. 950–1150 CE) in central Mexico (c. 96 km, or c. 60 miles, north of Mexico City), with whom the Maya, especially in Chichén Itzá, interacted. K'uk'ulcan has terrestrial and celestial elements, and since serpents shed their skin, the feathered serpent also signifies renewal and rebirth. This was not the first contact with central Mexico. The most well-known contact happened late in the fourth century CE when the "arrival of strangers" took place, and intruders, possibly from Teotihuacan (c. 50 km, or 31 miles, northwest of Mexico City), made their way through the western lowlands, eventually establishing a new king at Tikal. The Maya did not exist in isolation.

The Postclassic (post-900 CE) Maya wanted to get away from anything related to Classic Maya kings, who in the end could not convince ancestors and gods to end the prolonged droughts. They continued to build pyramid temples, but for K'uk'ulcan and other gods, not for kings like we see in the now empty southern lowland cities. Other Mexican influences include the *chacmool* (a stone sacrificial repository in the shape of a reclining feline or person), the lack of any royal portraiture, the growth of military societies, monumental columns, and noticeable localized styles (e.g., the Puuc style in the southwest part of the northern lowlands). Further de-emphasizing royal ideology, Postclassic kings no longer relied on the Long Count—a linear means of recording history—royal history, that is. Instead, they used the Short Count, which is a more cyclical, seasonal means of recording history that focused on renewal. It is comparable to July 4 (Short Count) versus July 4, 1776 (Long Count). Maya kings also ceased recording dynastic histories. The "silent" historic record made little difference to most Maya, who always had relied on their long-standing reciprocal or mutual relations with nonhumans and knowledge of seasonal practices from their forebears.

Their story continued—farmers farmed, elites displayed, and leaders led—until the early 1500s when another wave of newcomers came to the Americas, but this time, from the east across the Atlantic Ocean.

The Spanish Invasion and its Aftermath

Maya farmers lived dispersed in farming communities until the Spanish invaders forced them to congregate in villages (*congregación*) built around the *zocalo*, the hallmark of colonial cities. A *zocalo* is a plaza surrounded by a church, an administrative building, and a palace or official residence. This forced separation from their ancestors and way of life had physical and metaphysical repercussions still felt today. Concentrated people are easier to control. *Milpas*, however, were still spread out, which meant that Maya farmers had to add transportation costs getting to their *milpas*, not to mention what they owed to the Spanish in terms of tribute and labor demands.

It all began in 1502 during Christopher Columbus' fourth voyage to the Americas when explorers encountered Maya in their seafaring canoes.[10] The first face-to-face encounter took place in 1511 when the Maya captured shipwrecked Spanish soldiers who had landed on the east coast of the Yucatán. Even though these encounters were brief, the repercussions were deadly. Epidemic diseases including German measles, smallpox, influenza, bubonic plague, and typhoid were unknown in the Americas prior to the arrival of Europeans. Within a few hundred years, 90% of Indigenous Americans were wiped out. Until the arrival of Europeans, there were no epidemic diseases in the Americas because First Americans didn't domesticate many animals and there thus was little chance for the transmission of diseases between animals and humans (i.e., zoonosis).

Between 1515 and 1516, an epidemic disease, likely smallpox, spread in the northern lowlands. This epidemic probably explains why smaller cities like Saturday Creek and Barton Ramie to the south were depopulated in the early 1500s. Saturday Creek was a relatively small but bustling Maya city for about 2,100 years (c. 600 BCE to the early 1500s CE) due to its location over an aquifer on rich alluvium along the Belize River.[11] Bishop Diego de Landa, infamous for burning hundreds of Maya codices (books), wrote about a similar scourge in 1566: "A pestilence seized them, characterized by great pustules, which rotted their bodies with a great stench, so that the limbs fell to pieces in four or five days."[12]

Spanish inroads continued, and while Hernán Cortés focused on the conquest of the Aztec Empire between 1519 and 1521 in central Mexico, laying siege to their capital of Tenochtitlán, minor skirmishes between the Spanish and the Maya turned into a full onslaught beginning in 1523 when Pedro de Alvarado, aided by Kaqchikel Maya peoples, routed the K'iche' Maya in highland Guatemala.[13]

The Spanish faced many challenges. Between tropical conditions and diseases, dispersed farmers, and fierce resistance from the Maya, it took decades to establish a foothold. The Spanish capital of Mérida, for example, was not established until 1542. And the last Itzá Maya stronghold, Nojpetén (or Tayasal) on an island in Lake Petén Itzá in Guatemala, did not fall until 1697. Resistance continued for years, but in the end, the different fighting styles and weaponry, including guns, steel, horses, and attack dogs, were too much, especially when many Maya were dying of disease.

It wasn't enough to destroy Indigenous cities. Forced conversion to Catholicism followed. Spanish conquistadores built Catholic churches and cathedrals literally on top of razed temples. That is why one often hears about new discoveries in Mexico City being uncovered while workers are repairing sewage or subway systems. The Spanish literally built their new capital, Mexico City, over the Aztec capital, Tenochtitlán. Nearly every other major Spanish town was built over other destroyed cities. They also used pre-Columbian decorated blocks from destroyed temples, as if to say, "we are absorbing and replacing your belief system." It was always about the three G's: God, Glory, and Gold. But there is no gold in the Maya area; however, there is fertile land for the plantations or *haciendas* that sprang up all over that relied on forced Maya labor to grow maize, sugar cane, and henequin, as well as raise cattle.

Since the conquest, non-Maya have lived in the former British colony of Belize, including descendants of Spanish conquistadores, British adventurers, enslaved Africans (Creoles), and more recent migrants from Lebanon, China, the United Kingdom, and the United States. The Maya live in villages in the interior, and non-Maya live in towns along major roads. The Maya are marginalized, face discrimination, and typically have lower-paying jobs. Having driven to Belize from the United States several times, I never saw any Indigenous people owning large resorts, hotels, or stores in Mexico,

Guatemala, or Belize. There is a clear distinction of the haves (not Maya) and have-nots (Maya).

The long history of settler colonialism, where Indigenous peoples have little or no control over property, land, or the archaeological record, has resulted in the descendants of invaders and other foreigners controlling lands, sites, the government, and education.[14] For instance, Maya were not even allowed to perform traditional ceremonies in Tikal until 1996. Nor do Maya have a voice in politics. Even winning the Noble Peace Prize in 1992 did not help K'iche' Maya Rigoberta Menchu's presidential bid in Guatemala in 2011.

I cover political events in Guatemala to highlight the impacts of settler colonialism in the Maya world. But keep in mind that this story is all too common in other colonized places throughout the globe. Guatemala gained independence from Spain in 1821 after 300 years of colonial rule. In 1954, a right-wing coup funded and armed by the CIA installed a brutal authoritarian presidency against which several junior military officers revolted in 1960. This started a 36-year civil war that left about 200,000 Maya dead or missing.[15] Maya were lined up, shot and dumped into mass graves, especially between 1979 and 1984, when 91% of Maya massacres occurred. Things started to improve after 1984 when the violence deescalated and several democratic presidents were elected.

The civil war wasn't considered over until 1996 after a series of peace agreements brokered by the United Nations were signed. The powers that be also had to address the severe human rights violations. Many archaeologists and forensic anthropologists excavated mass graves and identified victims, returning the remains of loved ones to their families. Several thousand Maya still remain classified as "missing." In 2001, Juan Antonio's first year working with us, one of my crew chiefs told me near the end of our three-plus-month field season that "Juan Antonio has finally spoken to us. He said he has been in the Valley of Peace Village 12 years, and that he has not seen any of Belize for fear of the people." He lost his entire family to violence during the turbulent civil war in Guatemala and escaped to Belize to start life anew. He doesn't talk about what happened.

The story is the same elsewhere in the Maya area. In Mexico, however, the Zapatista Movement, or Ejército Zapatista de Liberación Nacional (EZLN), largely comprised of Tzeltal and Tzotzil Maya, emerged beginning in 1994 in Chiapas, Mexico, protesting injustices and racism. It is named after Emiliano Zapata, who early in the twentieth century campaigned for the restoration of village lands confiscated by *hacendados* (owners of *haciendas*), with the slogan *"tierra y libertad"* (land and liberty). Zapata was a leading figure in the Mexican Revolution, which began in 1910, and is one of the most revered national heroes of Mexico.

The fight continues today, in various forms and with varying degrees of success.

***

While the 500 years since the Spanish invasion has changed many things, one thing that has not been lost is Maya traditional knowledge. The Maya still hold on to the knowledge their forebears have been sharing for thousands of years, and it is this knowledge that can play a major role in the survival of our planet.

Chapter 3

Chahk, the Capricious Rain God

First the earth was created, the mountains and the valleys. The waterways were divided, their branches coursing among the mountains. Thus, the waters were divided, revealing the great mountains. For thus was the creation of the earth, created then by Heart of Sky and Heart of Earth, as they are called
POPOL VUH, *sixteenth century K'iche' Maya origin history*[1]

While the Maya appreciated, respected, and revered their gods, they also had to appeal for consistent rain—not the tropical storms that wash out seeds from the rich Earth or that ruin growing maize stalks and other crops. Nor the hurricanes that leave swaths of destruction. Yet it must be enough to nourish the crops throughout their growing cycle. In return, the gods required appeasement and gifts—the first fruits of harvest, sacrifice, and blood. But gods are capricious, and Chahk the Maya Rain God was no different. Too much rain. Not enough. Too early. Too late. Or just right. Chahk played a vital role because everything is rainfall dependent, and he must be appeased. Not surprisingly, the traditional Maya calendar is full of rain ceremonies—including Christian ones. They have been rainfall dependent from the beginning and continue to be so.

The power of the wet and dry seasons cannot be emphasized enough, and it highlights the skill and knowledge on which the ancestral Maya relied to adapt and thrive in such conditions. Each year the landscape transforms from a green garden to a green desert. Annual rainfall varies locally and regionally, ranging from 1,350 to 3,700 mm (53 to 146 inches). The beginning of the rainy season can vary by several weeks in any given area. Ninety percent of the rain falls during the seven-month rainy season from mid- to late June through January. As mentioned previously, much of the rain percolates through the porous limestone bedrock. Most farming activities take place during this season and involve circumnavigating swampy and flooded areas. Getting around can be challenging, especially at the height of the rainy season when rivers are turbulent with lots of debris (e.g., the Belize River carries debris from Guatemala). It is too much for many rivers at present for the few remaining hand-cranked ferries that still are in use. Until the all-weather Agripino Cawich Bridge opened in 2005 to replace the Young Gal ferry over the Belize River, the owners of Banana Bank Lodge, located on the north side of the river, kept vehicles on both sides that they reached via a boat; it was the only way to drive on the north and south sides of the river for *months* during the rainy season. The river was just too precarious, and long-distance travel was only practical during the dry season—to a point.

In the dry season the roads are so dry that dust clouds blind anyone driving behind you. You see thirsty animals, a yellowish tinge throughout the jungle, and low water levels. Rivers are low and navigable by canoe—but increasingly murky and disease-ridden, prime conditions to spread coliform bacteria (e.g., *Escherichia coli*). River ferries become immobile bridges. The days are brutally hot and humid. It is ironic that you can be in the middle of the verdant jungle yet die from dehydration if you don't know where to find water. Ancestral rural Maya flocked to cities. They needed water. They needed clean water.

Water quality, especially regarding drinking water, has always been a concern in the hot and humid tropics where c. 43% of the world's current population resides, as well as 80% of the world's terrestrial biodiversity.[2] The tropical belt, currently 23.5 degrees north and south of the equator, as mentioned earlier, covers c. 40% of the Earth's surface and will likely expand because of global climate change.[3] People also need to drink more water (2–3 liters per day more) than those living in temperate areas since you can lose up to 10 liters of water per day sweating. Cooking, bathing, manufacturing (e.g., ceramics, plaster) and other activities also require water, totaling 20–50 liters per day per person according to the United Nations.[4]

Water is valuable.

Today people living in Central America are facing changing weather patterns related to global climate change: Will the heat and humidity get worse? Yes. Will tropical storms and hurricanes intensify in number and magnitude? It looks like this is the case. Unless dealt with, climate change will impact water supplies, especially water quality. Will we see more climate refugees attempting to escape increasingly precarious conditions, just like what happened in the Terminal Classic (c. 800–900 CE) when about 90% of the Maya emigrated out of the interior southern lowlands? Yes.

Curtains of Rain

At the Choc house in the Valley of Peace Village one morning in June 2022 when it was raining hard, Cleofo's 4-year-old granddaughter, Loui (short for Nathania Louisa Choc) declared, the sun is fighting.[5]

Walls of falling water. That's what localized storms look like on the horizon. The demarcation is striking. You could have one foot on dry ground, and the other in pouring rain. You can drive into a curtain of rain as it hits the hood of your vehicle first, followed quickly by reaching the windshield. One minute it is sunny, and the next breezy. Whenever there is a nice breeze, it is smart to look up to the skies to see if there are dark clouds. Or you hear it, as a sheet of rain less than a minute or so away. And then you see it—a sheet of rain coming right at you.

Rains start usually in mid- to late June and pick up soon after. Archaeologists excavate during the dry season, usually in mid-May when the semester is done, when one has about six to seven weeks before the rains derail excavation or survey plans. Archaeologists can't excavate in the rain, and 4-wheel-drive trucks can only drive so far when dirt roads are scarred with deep, muddy ruts or washed out altogether. Limestone roads also are slippery when wet. And when the ruts are too deep, getting stuck and unstuck is time-consuming and exhausting. Archaeologists use engine-powered winches, shovels, machetes to cut branches with leaves to fill ruts, and other means to do whatever it takes to get unstuck. Ironically, vehicles can still get stuck at the end of the dry season if roads have lots of clay shaded by trees. In such a situation, the clays remain wet and difficult, if not impossible, to traverse. Clearly, the Maya have been dealing with these same issues for millennia—and quite well.

The rains transform everything. It is a time of renewal; the rains nourish flora and fauna, and the jungle turns green. It is also the agricultural-intensive period, so there is always the risk of flooding, erosion, and crop damage. Interestingly, maize is a relatively thirsty plant compared to, for example, millet, sorghum, or manioc.

The dry season promotes the fecundity and proliferation of pests that eat maize and other crops, whereas rainfall enhances the growth of fungi that kill off pests (e.g., spider mites and army worms). But rain and water can carry disease, as illustrated by the maize fungus spread by rain and wind, *Ustilago maydis*, also known as corn smut. Also, drought or improper storage of maize promotes the formation of aflatoxin, a liver carcinogen.

Predicting when the rains will begin is challenging. If farmers plant too soon, the seeds will rot, or ants will eat out the soft parts. If they plant too late, the seeds will not germinate. When the rainy season begins, life emerges amazingly fast. Driving past fields every day you can see maize stalks grow from one day to the next. June bugs, or large, flying beetles (genus *Phyllophaga*), come out at night in droves after the first rains, crashing into any lights and anyone too close. Flying termite queens (genus *Reticulitermes*) emerge from nests in the thousands all day for a week or so looking for a home to start their own colony—getting into your food, water, hair, clothes, and just everywhere. Extra protein never hurt anyone.

Heavy rains wash out freshly planted seeds. Rivers and creeks overrun and flood large areas and make roads and paths impassable. And then there is the problem of long-term food storage in the humid tropics, where foods and organics decay rapidly. To address this issue, Maya farmers turn maize ears downward while they are still in the *milpas* to prevent damage from rain and pests such as birds and coatimundi (a raccoon-like omnivore) and collect them when they need more.

Bajos (seasonal wetlands or swamps) remain flooded and impassable throughout the rainy season and even parts of the dry season. Hail can appear out of nowhere, even during the dry season. The hurricane season usually takes place anywhere from July through October. But hurricanes are occurring earlier and earlier—likely the result of global climate change. In 2008, for example, Hurricane Arthur struck in late May. Hurricanes are mentioned in Classic period inscriptions, represented by the sky glyph surrounded by wind glyphs—a "hurricane of four winds," according to archaeologist and epigrapher Stephen Houston. Yet ancient Maya buildings are still standing. This fact was brought home in the aftermath of the October 24, 2010, Hurricane Richard when part of the 648 km^2 (250 square miles) Yalbac Ranch property was ground zero and Maya sites were in the direct path of the hurricane, including the ancient Maya city of Yalbac. It has a reservoir, large plazas for public events, administrative buildings, a ball court, six pyramid temples, and a royal acropolis over 20 m (66 ft) tall (see Figure 4.3). Viewing the aftermath of the destructive hurricane was difficult for everyone—locals, loggers, archaeologists, and anyone else who bore witness to the hurricane's wrath. I could not help but wonder how the ancestral Maya responded to extreme weather events.

As briefly mentioned earlier, in February 2011 I conducted a posthurricane assessment of Yalbac at the request of the Belize Institute of Archaeology.[6] Massive trees were strewn all about. However, there was minimal damage to Yalbac's monumental buildings constructed of cut limestone blocks and plaster mortar. Even though the Maya abandoned Yalbac over a thousand years ago, it withstood the devastating hurricane and subsequent wildfires during the 2011 dry season. The vaulted roofs the Maya had sealed to protect against leaking worked. However, the damage to modern buildings, roads, infrastructure, and vehicles was noticeable, including at Banana Bank Lodge. The hurricane tore off the thatch roof from the large dining hall, 24-m (80 ft)-tall (and taller) trees were broken in half, the exquisite orchid house was wiped out as if it never existed, and nearly every surface was under water.

Rains impact everyone and everything. Archaeologists lose many days of field work due to rain. Outdoor activities are out of the question, too. Traditional Maya built their houses on slight rises or built platforms to keep the water out. And their thatch roofs are waterproof for about 20 years before they have to be replaced (more about houses in Chapter 7).

Sometimes the rainy season begins with tropical storms, several right in a row. We're talking buckets of blinding rain and wind. All-weather dirt roads are washed out; parched and hardened dirt roads are unable to absorb all the rainwater. The baked roads course with rivulets, which cut small ravines into the surface like a nest of slithering snakes—sometimes creating gaps too large to drive over. River ferries become useless.

Today, hurricanes and tropical storms often take out electrical power, wash out roads and bridges, topple buildings, and destroy crops. In fact, the Belize Government moved its capital inland from the coastal city of Belize City in the 1970s to escape the continual rebuilding after particularly harsh hurricane seasons. Belmopan emerged from a low-lying savanna landscape to become the new capital, and embassies soon followed. The Classic Maya did not need to move their cities. There are no known Maya sites in the Belize City area or other areas susceptible to flooding. For example, in 1999 when mapping the small Maya city of Saturday Creek, we noticed that some of the lower river terraces lacked any settlement, whereas the higher ones had lots of variously sized mounds. Upon our return in 2001 the lower terraces were under water. The Maya knew where to build and where not to build.

A Green Desert

Most everyone in Belize can't wait until the rainy season starts—to plant, to cool off, and to fill desiccated creeks and lakes. Not archaeologists. The green desert is perfect for conducting archaeology. However, the dry season has its own challenges. "Dry" is a relative concept due to the extreme humidity and the unexpected tropical storms that appear out of nowhere, especially at the end of the dry season in late May to mid-June. I should note that since archaeologists only work in the dry season, our models and ideas might be a little biased.

The five-month dry season is a challenge today, as it was in the past. Surface water decreases throughout the dry season due to high evaporation rates because of the intense heat. High humidity also is an issue because of how it translates into high heat indexes. For example, 90 degrees Fahrenheit plus 80% humidity means a heat index of 113 degrees! The summers of 2016 and 2022 were particularly hot (the heat index reached 120 degrees), and we were excavating house mounds on Spanish Lookout property—a treeless landscape. In 2018 while we were trying to find a way to the Motmot Sinkhole from the cliff above a Cara Blanca *cenote* through thick postfire secondary jungle growth, it was suffocating, with nary a breeze. I ran out of water, even though I started out with two full canteens, and suffered from heat exhaustion. Water is vital.

The water table is too deep to reach in many parts of the southern lowlands (100+ m, or 328 ft), as the University of Pennsylvania and Guatemalan Government archaeologists found out in the 1960s at Tikal in the Petén district in Guatemala. After having drilled up to 180 m (591 ft) for wells without hitting water, they ended up having to rely on ancient reservoirs. The amazing thing is that these reservoirs had not been maintained for over a thousand years. Yet they provided not only enough water for drinking, but enough for reconstructing and consolidating the monumental architecture. However, in 2016 at Tikal, it was the first time one of the few remaining reservoirs had dried up. The drought was so bad that park staff put out water troughs for animals.

Today, many ancient Maya reservoirs at the hundreds of cities have dried up since they have not been maintained since the ninth century CE urban diaspora, but not all of them. Diane and Arlen Chase at Caracol also relied on ancient reservoirs for drinking water (they added chlorine to make it potable), as well as for bathing, cooking, and consolidation (to make plaster and mortar). It would be interesting and useful if people could bring these reservoirs back to life, especially in the face of worsening droughts in Central America according to the United Nations World Food Program (https://www.wfpusa.org/emergencies/dry-corridor/). During the nineteenth century economic expansion in the Yucatán, Mexico, hacienda owners resurrected ancient Maya water features (e.g., filtration wells) in the Puuc hills in the northern lowlands.[7] Also, anthropologists are working with local Maya to resurrect canals at the Maya site of Cauich on the northwest coast of the Yucatán in Campeche, as well as those near Calakmul.[8]

There is a three-month block of time in the five-month dry season when it does not rain at all, not a drop. Six months after the October 2010 hurricane toppled too many trees to count, they dried out. Humidity was no match for the intensity of the tropical sun beating down on the exposed, dead trees. On the Yalbac Ranch property, delays in getting permission from the Belize Government forestry office to collect fallen trees smaller than the legal minimum girth loggers usually are allowed to remove meant that the fallen, dead trees created a huge tinder box. What followed was what everyone dreaded—massive wildfires. By May 2011, large swaths of the same forest felled by the gods were burning. Smoke, haze, and floating ash spread as far as the eye could see and made living and working in Belize difficult, unpleasant, and unhealthy. Thousands of hectares burned on the Yalbac Ranch property. Animals ran away from the fires, trying to escape.

Witnessing the hurricane's wrath was difficult enough. Witnessing the erasure by fire of what was left was heartbreaking. The scale of the destruction was far-reaching, especially for jungle life. With so many fruit trees struck down or going up in smoke, howler and spider monkeys and other forest inhabitants who rely on fruits, blossoms, leaves, and nuts for their survival, could have starved. Cleofo informed us that deer are better off than other animals because they have more options of what they eat and that howler and spider monkeys and many other fauna that escaped the fires moved north to a fire-free jungle forest. Flora and smaller animals were not so lucky. For several years after the fires, the jungle was so quiet. Too quiet.

The ancestral Maya adapted to this unpredictable seasonal life and created their own water supplies that not only lasted throughout the dry season but remained drinkable.

Maya Self-Cleaning Reservoirs

While in the Jaguar Inn bar, one of the few on-site hotels, on my second visit to Tikal in 1989 I was listening to a conversation between archaeologist Anabel Ford and Felipe Lanza, who at the time ran Tikal's Flora and Fauna center and had worked at Tikal National Park for 32 years. They were speaking in Spanish, discussing reservoirs and *aguadas* (natural depressions often lined with clay to retain rainwater). Felipe told Anabel that he knew every *aguada* and reservoir in the park—103 of them—and that they all have Maya sites associated with them. The larger they are, or the more of them there are in any given area, the larger the site or city. This pattern makes complete sense.

The ancestral Maya created their own urban water supplies, namely, artificial reservoirs. Reservoirs, however, are moot if their water is not clean enough to drink. Stagnant water is undrinkable and serves as breeding grounds for mosquitoes and water-borne diseases and parasites. Mosquitoes spread malaria, West Nile virus, dengue, yellow fever, and others. Interestingly, before the Spanish arrived in the Americas in the late 1400s, mosquitoes did not carry these diseases.[9] That said, depending on where they lived, pre-Hispanic Native Americans still suffered from a myriad of diseases—chagas, amoebic dysentery, leishmaniasis, treponemiasis, tuberculosis, pertussis, giardia, rabies, tularemia, herpes, hepatitis, and poliomyelitis.[10]

How did the Maya maintain water quality, especially since standing water becomes stagnant due to algal growth resulting from the build-up of nutrients like phosphorus and nitrogen (eutrophication)?[11] I was determined to find out. When I was a Chancellor's Postdoctoral Fellow at the University of California at Berkeley (1995–1996), I had several conversations with civil engineers who were exploring nonchemical methods of cleaning water in California as an alternative to water treatment plants. They were interested to know how the urban Maya dealt with potential human waste contamination in view of the porous, karstic landscape and their reliance on artificial reservoirs.

In brief, the Maya relied on a diverse array of aquatic plants and other biota to maintain water quality in the same manner as do constructed wetlands (CWs) today that use "natural processes involving wetland vegetation, soils, and their associated microbial assemblages to improve water quality."[12] In other words, CWs are self-cleaning thanks to certain macrophytic plants (growing in or near water) that decrease eutrophication by consuming nutrients like nitrogen and phosphorus. Macrophytic plants include water hyacinths (*Pontederia crassipes*), cattails (*Typha domingensis*), rushes (*Juncus* spp.), bulrushes (*Scirpus* spp.), duckweed (*Lemna* spp.), and reeds (e.g., *Phragmites australis*). Further, microbial biofilms nourished by decomposing plant matter break down nutrients. Also involved in the cleaning process are hydrophytic plants (roots growing in the water) such as pondweeds (e.g., *Potamogeton*). Further, different kinds of zooplankton in CWs prey on pathogens (e.g., rotifers, protozoa, and nematodes), and different kinds of bacteria (e.g., *Pseudomonas*, *Alcaligenes*, and *Bacillus*) assist to denitrify water and consume harmful microorganisms (e.g., parasites).

Maya reservoirs also likely contained a myriad of macrophytic and hydrophytic plants based on what we find today in Central American wetlands, including edible and medicinal plants (e.g., *Rauvolfia* spp., *Lacmellea* spp., *Ficus* spp., *Typha domingensis*, *Cyperus esculentus*, *Pachira aquatica*, *Pontederia cordata*, and *Sagittaria* spp.), bamboo (*Merostachys* spp.) to make fish spears and other implements, reeds for basketry (e.g., *Phragmites*), and others.[13] Cyperaceae (sedge) and other aquatic plants have been found in sediment cores extracted from ancient Maya reservoirs. Maya reservoirs also supported diverse aquatic life such as turtles (e.g., slider), crustaceans (e.g., blue crab and shrimp), eels, mollusks, snails, and fish (e.g., cichlids, tropical gar, and catfish). Fish eat pesky insects and their feces and other bottom debris, which the Maya would have had to dredge every few years and which would have served as an excellent fertilizer for urban gardens, fields, and orchards. The Maya also would have replaced and replenished aquatic plants that they likely used as fertilizer as well since they would have been saturated with nitrogen, phosphorus, and other nutrients.

Maya reservoirs at the hundreds of cities supplied millions of people for over 1,000 years. But how did the Maya know when the water was clean enough to drink? By the presence of water lilies

(*Nymphaea ampla*), a hydrophytic plant, on reservoir surfaces. Previously, archaeologists had assumed that water lilies played a role in keeping water clean not only because of their prevalence in the iconography and on water bodies at present, but because all plants take up nutrients like nitrogen and phosphorus. However, water lilies are quite sensitive. In fact, in the library devoted to water at UC Berkeley (Water Resources Center Archives[14]), I found hundreds of references on hyacinths (macrophytic) but could only find four on *N. ampla*, and they all had to do with how to build your own water lily pond—which can be challenging. Why? Because *N. ampla* can only grow in water that is not too deep (1–3 m, or 3–10 ft), does not have too much calcium (and other minerals) or too many algae, and is not too acidic. They can only grow in still, slightly alkaline water. In short, they can only grow in *clean water*. So, when you see water lilies on the water's surface, you know the water is clean enough to drink.

The Maya lined reservoirs with natural sealants like clay. They then added soil to support water lily roots and other aquatic life; doing so also helped stabilize pH levels and counteracted the toxicity from released gases from decomposing plants like methane, ethylene, and phenols. Water lilies have additional benefits. Their pads restrict the passage of light, which inhibits algae growth and evaporation, keeps water cool, and provides cover for predators of mosquitoes and their larvae, such as fish and turtles. Archaeologists have found *N. ampla* pollen from several ancient reservoirs, for example, at Xultun (c. 10 km, or 6.2 miles, south of San Bartolo, Guatemala) and near Nakbe in Guatemala.

Water lilies indicate potable water. Clean water is associated with Maya kings (more about the relationship between water and royal power in Chapter 8).

The Power of Reservoirs

Archaeologists find reservoirs difficult to date since they are rarely explored, though excavations and coring at some of Tikal's reservoirs are beginning to reveal their secrets. Archaeologist Vernon Scarborough and his team have shown that Tikal's central plazas fed water into a series of three reservoirs separated by coffer dams, including one 76 m (250+ ft) across and 9 m (30 ft) high, the largest one built by the Maya. They also released gray water to downslope and *bajo* margin reservoirs and *aguadas* for orchards, gardens, urban fields, fishponds, construction projects (e.g., to make plaster and mortar), and to attract game and waterfowl (e.g., deer, tapir, peccaries, herons, limpkins, cormorants, and ducks). Tikal's layout basically revolves around water diversion, capture, and storage. The city *was* a water management system.

The ancestral Maya began capturing and containing water with still-water systems by at least c. 400 BCE with wetland reclamation and gravity-fed depression filling reservoirs as seen at El Mirador in the Petén, Guatemala. In fact, the Maya built water systems even before they built pyramid temples. Some of the earliest and largest reservoirs failed, perhaps due to silt build-up and second century droughts, which was the case at the large Preclassic cities of El Mirador, Nakbe c. 15 km (9.3 miles) to the southeast (600 BCE to 150 CE), and others in the Mirador Basin that were abandoned by 150 CE.[15] El Mirador still holds the record for the largest temple complex in the Americas, today known as La Danta, which towers over 70 m (230+ ft) overlooking a city nearly 10 km^2 (6 square miles) in size. All monumental constructions were slathered with thick layers of plaster, which would have required lots of wood, limestone, and water to manufacture. Between the needs of plaster production, fuel, and agricultural fields, El Mirador's inhabitants deforested the surrounding areas, which led to erosion and silt build-up.

In contrast, the Maya did not abandon Tikal during the same time period, but the second century CE droughts led to decreasing water quality (e.g., too much nitrogen and phosphorus promoted cyanobacteria that could have resulted in toxic algal blooms) as evidenced at the Palace and Temple Reservoirs of Tikal.[16] For most cities, however, the reason we do not see a major urban diaspora like we do at the end of the Classic period comes down to the fact that in the Late Preclassic (c. 250 BCE to 250 CE) there were fewer people, less land use, and less overuse of resources. In other words, the Maya could more easily cope with climate instability.[17] Following the Late Preclassic droughts, the Maya did not have to abandon southern lowland cities for another 700 years because they learned from their mistakes.

At the dawn of the Early Classic in 250 CE, the Maya engineered increasingly larger and more sophisticated reservoirs to capture and store greater amounts of water. Quarrying of reservoirs and associated features not only provided construction materials to build monumental buildings, but also influenced urban layout along with topographic and cosmological factors. Some of the earliest public iconography depicts aquatic elements such as the Waterlily Jaguar God. By 500 CE, the reservoirs at Tikal had grown to cover over 12 hectares (30+ acres) and could hold over 900,000 m^3 (c. 240+ million gallons) of water in a year with typical rainfall (c. 1,500 mm, or 60 inches) and could supply tens of thousands of thirsty subjects over the long dry season (millions of gallons, where 1 million gallons = 3,785+ m^3) (Figure 3.1).

Water systems reached their height in sophistication in the Late Classic (c. 600–800 CE) via elevated stream-damming reservoirs that captured and stored rainwater but diverted excess water aided by sealants (e.g., rammed earth, clay, and limestone slabs), dams and berms, sluices, spillways, channels, switching stations, and filtration systems (e.g., the Maya transported quartz sand to Tikal from up to 39 km, or 24 miles, distant, which was used to filter water). This would have supplied the city of 60,000 to 80,000 people with plenty of water through the dry season, given that 1 m^3 equals 1,000 liters for five months, where 80,000 people would have required 400,000,000 liters or 400,000 m^3 (105,668,821 gallons).

One potential problem with the porous limestone bedrock is seepage of, for example, human waste. The Maya generally did not

FIGURE 3.1 *Lidar map of the urban core of Tikal with several reservoirs highlighted*

Source: Modified by Bryan Lin from Figure 2 in L. J. Lucero. 2023. Ancient Maya Reservoirs, Constructed Wetlands, and Future Water Needs. *Proceedings of the National Academy of Sciences* 120:e2306870120. doi: 10.1073/pnas.2306870120. Used with permission.

build structures near reservoirs. For example, David Lentz and colleagues analyzed environmental DNA (eDNA) from reservoir sediment cores at Tikal and found that reservoirs were ringed with tropical forest vegetation, including large trees (e.g., *Brosimum alicastrum*, or breadfruit tree, *Ficus* spp.), small trees (e.g., *Lacmellea* spp.), palm trees (e.g., *Cryosophila staurocantha*), shrubs (e.g., *Rauvolfia* spp.), and vines (e.g., *Cynanchum* spp.).[18] Disturbance plants (weeds and secondary growth) only made an appearance in the later part of the Late Classic and in the Terminal Classic, as did increasing contamination, including cyanobacteria from algal blooms and phosphate (PO_4^{3-}) from fecal matter and food waste—that is, when the prolonged droughts struck and things started to fall apart.

Urban dwellers drank, cooked, washed, produced pottery, and made the plaster that adorned temples and sealed plaza floors to funnel rain into the reservoirs.

***

The tropical forest, jungle life, and seasonal weather shaped Maya life, agriculture, and cities. The stage now is set to address the essential question of how the Maya successfully coexisted with nonhumans for millennia. The short and simple answer is because of their inclusive worldview, one that guided their daily existence and engagement with the nonhuman world.

Chapter 4

The Maya Inclusive Worldview

THEN were conceived the animals of the mountains, the guardians of the forest.
  POPOL VUH, *sixteenth century K'iche' Maya origin history*[1]

Missionization in the Maya world began 500 years ago with the arrival of the Spanish and continues today. This onslaught on the Maya inclusive worldview has eroded much of its power—and potential. Syncretism, however, still provides glimpses of the Maya worldview; for example, statues of Catholic saints are clothed and stored in traditional Maya ways. The symbol of the ceiba tree, the tree of life with its horizontal branches, looks just like the Christian cross. There are many such examples. Here, I focus on how the Maya continue to engage with the tropical world via vestiges of their worldview and their traditional knowledge of forest collaboration, weather and seasonal patterns, and agricultural practices.

My knowledge of the Maya nonanthropocentric, inclusive worldview comes from ethnographies, ethnohistories, Maya stories, poems and origin histories, Classic period inscriptions, and the archaeological record. I interpret the landscape created or inscribed by Maya gods and how the Maya engaged with this charged landscape in addition to that created by people through their houses, inscriptions, iconography, monumental architecture, and household and ceremonial items. Though I piece together different lines of evidence about the ancient Maya worldview, I typically use the present tense for flow and because it best reflects their cyclical view of the world, which merges the past, present, and future. I follow the lead of the Maya authors of the *Popol Vuh*, the sixteenth century K'iche' version of their origin history, who wrote in the same manner.

The Maya live their worldview daily, including via frequent interactions with other entities, particularly openings in the Earth like caves and water bodies that are portals to the Underworld, where they continue to conduct ceremonies and leave offerings for ancestors and gods, which today include Christian ones (God, Jesus Christ, the Virgin Mary, and saints). Portals weave throughout Maya life in complex and multidimensional ways. For instance, they deposit dangerously imbued entities or beings in portals—like accused and punished witches. There are so many levels of engagement—but it works. It's all about checks and balances guided by their inclusive worldview.

The Maya Way

The Maya inclusive worldview is one where things, humans, animals, land, water, and other entities exist on the same plane, where each plays a role in maintaining themselves and the world.[2] Each is animated and has a *ch'ulel* (soul). Each embodies *k'uh*—for lack of a better term, a sacred aspect, which "inhabited all things in the universe—rocks, trees, humans and all living beings."[3] Entities coexist. Each connects or communicates with other beings through their souls. As an example, when the Tzotzil Maya of Zinacantán, Chiapas, Mexico, perform cargo rituals,[4] "Three shots of rum are added to make the squash turn out well. It is believed that the 'soul' of the pot receives the 'soul' of the rum, and, in turn, blesses the cooking squash and prevents the cooking vessel from breaking."[5]

FIGURE 4.1 *Close-up of traditional Maya cotton* huipil *woven with a backstrap loom*

Source: Photos by the author.

This worldview is expressed in the *huipil* (traditional Maya blouse) shown in Figure 4.1, which was made by a woman of the Jolom Mayaetik coop of weavers in San Andrés Larrainzar, Chiapas, Mexico, as part of Weaving for Justice, a nonprofit organization based in Las Cruces, New Mexico. As stated on their website,[6]

> The cosmogram depicts a quartered universe moving through time, uniting Earth and Sky. It also charts the path of the Sun, a principal Mayan deity. The cosmogram appears in various versions; in each version viewers look straight down on the world from just above the highest point of the heavens. Five diamond designs mark the four cardinal directions, and the central diamond may stand for the nadir, the lowest point under Earth where Sun passes at midnight on its circle back to the east.

The Maya live their worldview as fellow beings or entities. There are no Cartesian dichotomies—no sacred/secular, natural/supernatural, or animated/inanimate. English words cannot convey this enmeshed existence; we have no terms for culture/nature, death/life, destruction/creation, sacred/secular, and other dichotomies. So, I convey the intricacies of their worldview the best I can using an anthropocentric language that differs from many Indigenous languages where family terms are used to refer to nonhumans to highlight their kinship.[7]

The Maya live in a world where all entities—plants, animals, soils, water, and others—are part of the chain of life, similar to what early environmentalist Aldo Leopold argued for our ecosystem being "a tangle of chains so complex as to seem disorderly, but when carefully examined the tangle is seen to be a highly organized structure. Its functioning depends on the cooperation and competition of all its diverse links."[8] A change in one link has reverberations throughout.

The Maya work with nature, not against it. They do not attempt to dominate or control it, a fact reflected in their ergative languages (i.e., nonliving subjects act—e.g., the building collapsed) that also de-emphasize "I" and emphasize "we." "We" includes clouds, plants, rivers, mountains, animals, and other entities.[9] Nor are there terms for "religion" or "nature." The closest approximation for "religion" includes *Okol k'u*, "to enter God" in Yucatec Mayan, and *Utz Xanbal*, conveying "a (sacred) righteous way of life" in Tzotzil. Instead of "nature," the Maya use terms including "earth," "forest," "land," "world," "face of the earth," "soil-earth-world," and "universe," according to Mayanist Harri Kettunen.

The closest concept to dualism, if one can call it that, is their concept of complementary opposition; for example, portals have both dangerous and sacred aspects. It is comparable to the two sides of a transparent coin. That said, complementary opposition still embodies multiple aspects. All Maya have a co-essence or *way* (animal spirit companion) with whom they share a *ch'ulel*. Domesticated animals are never *way*, only wild ones, who live with and are taken care of by ancestors in forest *wits* (lineage mountains). *Way* bonds people and the forest, merging them, if you will.

The Maya acknowledge the need to consume other entities to survive, but under the guise of mutual respect, responsibilities, and obligations. It is all about balance and not overusing "resources." In the Yucatec Maya village of Chan Kom in the Yucatán, for instance, plants and animals who live near or in certain *cenotes* are sacred and cannot be killed. The Maya enter the jungle to hunt and collect food, medicinal plants, and construction materials. Simultaneously, they spiritually enter the domain of their animal spirit companions. Forest spirits, such as *kuilob kaaxob* among the Maya of Chan Kom, ensure that people do not take more than they need. Deities must grant their permission before Maya can hunt, fish, fell trees, or extract limestone and chert. In addition to permission, the performance of rituals also is required to give thanks to animals, trees, plants, and other nonhumans who give up their life. If Maya disrespect fellow guardians, ancestors make them pay through their *way*; if *way* become ill, so too do their humans. Everybody watches out for everybody else—that's how this system works. For instance, for the Lacandon Maya in Chiapas, Mexico, when they place the bones of hunted animals inside caves, they are not disposing of them. Rather, they are ceremonially depositing the bones as the first step in their regeneration by the Lord of the Animals.[10]

Maya lives are part of a continuous cycle in which death begets life and destruction begets creation, quite different from the linear system we espouse. Their cyclical view of life focuses on renewal—just like the seasons or daily rotation of the sun and moon. Then there's *ch'ulel*. It is like oxygen. We can't see, taste, smell, or feel it. But it is all around us and gives us life. Oxygen is recycled, as are souls. In fact, Maya newborns are referred to as *k'exol* or "replacements" of their ancestors. In most Mayan languages, people often refer to maize kernels as "little skulls," especially those used to reseed by entering the Underworld and later emerging through Earth to grow and nourish people and animals alike.[11] This is yet another example of the continuous cycle of death and renewal.

Death begets life. Beautiful orchids live off the dead—to be exact, decomposing vegetation. In the forest you often come upon large, toppled trees on the jungle floor from which are growing orchids, including the national flower of Belize, the Black Orchid; the dead tree nourishes this exquisite flower.

This cyclical world is like a continuously moving spiral through space and time, without a beginning or an end, where entities are renewed or reborn. The Maya live in a universe where the Upperworld has 13 segments, and the Underworld has nine. In between is where humans and nonhumans reside on the Earth' surface, depicted as the back of a crocodile or turtle floating on the primordial sea.

Gods also show how seemingly straightforward Maya concepts are instead complex and multidimensional. Chahk the Rain God, as I have shown, is not just about the rain; he is responsible for droughts, hurricanes, tropical storms, and too little rain. Michael D. Coe detailed how some gods have up to four different aspects related to the cardinal directions and associated colors.[12] Some also have a counterpart of the opposite sex. Some have young and old aspects or fleshed and fleshless ones. Finally, some have enmeshed animal-human aspects. In *The Book of Chilam Balam of Chumayel*, written the late eighteenth century, the Yucatec Maya refer to Oxlahun-ti-ku, or Thirteen Gods that signify the 13 realms of the Upperworld, and Bolon-ti-ku, or Nine Gods that signify the nine realms of the Underworld. Each, however, was probably seen as a single god.[13] And I am sure Maya gods had other aspects we are unaware of, perhaps along the lines of traditional Polynesian and other societies where the "gods are respectively manifest in colors; directions; days of the week; periods of the day; natural and inorganic phenomena, such as thunder, light, seawater, and so on; plants; animals; seasons; certain smells; cloud formations; a particular number; birdsongs; and more."[14]

Figure 4.2 shows a well-known unprovenienced (i.e., looted) Classic Maya dish that embodies the three realms of the Maya world. It shows the Hero Twins "watering" their father, the Maize God, who emerges from the turtle carapace reborn after having been killed by the Death Gods while playing a ball game in a ball court in

FIGURE 4.2 *Maya ceramic dish showing the Maize God being reborn*
Sources: Photo by Justin Kerr (K1892). Used with permission. Justin Kerr Maya archive, Dumbarton Oaks, Trustees for Harvard University, Washington, DC.

the Underworld. The painted scene depicts the watering, growth, and emergence of, well, everything. Newborns emerge from the wombs of their mothers. Maize erupts through the Earth's fertile soils and grows. The sun emerges from portals since the Hero Twins later become the sun and the moon. Death is necessary for the renewal of life. Everything is connected.

The archaeological record is replete with evidence of these renewal cycles. The stratigraphy archaeologists reveal, that is, repeated layers of floors and ballasts (floor supports), walls, doorways, and ceremonial deposits, reflects centuries upon centuries of renewal via burials, de-animation deposits, rebuilding, and animation caches. For instance, the multilayered house mounds enfold generations upon generations of family histories created by burying deceased family members, de-animating houses (i.e., releasing their *ch'ulel*) via rites, razing and burning everything, and rebuilding and renewing through animation rites in the same place, again and again (more about this in Chapter 7). Ceremonies accompany every stage of household life. Archaeologists find evidence for this renewal at all scales—from the smallest house to the largest palaces and temples, created by ceremonies that are essential to thank gods, ancestors, and fellow guardians.

The Maya world is perceived via its four cardinal directions and its center (i.e., a quincunx). Different birds, trees, colors, and deities are associated with the five points. East/red symbolizes the sun being born each day and renewal. West/black symbolizes the death of the sun each night. North/white symbolizes the rainy season. South/yellow symbolizes the sun and dry season. Center/blue-green symbolizes the *axis mundi*. One can't exist without the others.

The four limbs of people and animals signify the cardinal directions as well, and the heart literally and figuratively is the center. Bodies are of the same elements—Earth (body), wind (breath), and water (blood). The Tzotzil term for red tree resin and blood are the same, and the word for corn silk and hair are the same. The different varieties of maize come in the colors associated with the four cardinal directions—red, black, yellow, and white. For houses and other structures, the area underneath the roof signifies the sky; the walls, the mountains; and the floor, the surface of the Earth. But things are never so straightforward; there are always overlapping schemas. The house, for instance, also is referred to in terms used for the body—walls are the stomach; the foundation, the foot; the corner, an ear; and so forth. And ears of maize are referred to as "heads."

Archaeologists find evidence of this enmeshed cosmos, such as lip-to-lip vessels that are found in burials and ceremonial caches. They consist of two dishes, one inverted over the other and resting on the lower vessel—literally lip to lip. They often contain stones and jade signifying the Earth, as well as marine shell and nine chert flakes signifying the nine sectors of the watery Underworld. Chert is created by lightning and was also used to start fires for the gods. The space underneath the inverted vessel is the sky and Upperworld. At Yalbac in central Belize, excavations revealed a lip-to-lip funerary offering from a sidewall of a looter's trench at a small pyramid temple (Temple 3B, 20 × 20 m and 6 m tall, or 66 × 66 ft and 20 ft tall) (Figure 4.3). The inverted vessel on top is black, signifying west and

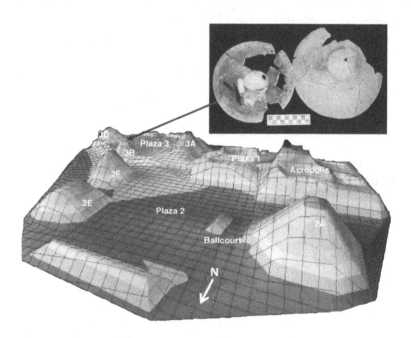

FIGURE 4.3 *Lip-to-lip funerary offering exposed in a looter's trench, Temple 3B, Yalbac, Belize*

Source: Courtesy of VOPA.

death. The bottom dish is red, signifying east and rebirth. Inside, the Maya had placed two freshwater *Pomacea* shells (water) with drilled spheres, a carved boar tusk (forest), and two thin black obsidian blades (Earth and/or death).

Another example of the enmeshed cosmos are pyramid temples—ubiquitous in every city. The Tzotzil Mayan word *wits* translates as "lineage mountain" (where ancestors reside) and "pyramid temple." Temples do not represent ancestral mountains. They are not replicas. They *are* ancestral mountains. Ancestors and gods reside in *wits* in the forest, while the Maya performed ceremonies and buried their deceased kings in urban *wits*. Ancient Maya pyramid temples often have nine terraces signifying the nine Underworld realms, while temple doorways signify portals. For example, the Temple of the Inscriptions in the major city of Palenque, Mexico, has nine terraces (see Figure 9.2), where the sons of its most powerful king, K'inich Janaab' Pakal (603–683 CE) interred their father at the base of a steep flight of stairs deep inside the *wits*. You can only access the tomb through entering a *ch'e'n* (portal or cave) via a staircase that goes deep into the *wits*.

Everything is connected—and coexists.

Impacts of an Inclusive Worldview

The Maya were excellent astronomers, not for the sake of understanding the nature of stars and the sun, but for astrological purposes. The Maya interweave their life histories with heavenly bodies through their calendrical system, which is part of their daily life from the moment they are born to the day they die, when their souls are recycled and it begins anew. Their life histories are connected to the Upperworld via the stars, the moon, the sun, and shooting stars.

The sacred almanac *Tzolk'in* ("order of days") consists of 20 day names meshed with 13 day numbers (260 days total—the same amount of time it takes for human gestation). The solar calendar, *Haab*, consists of 18 named months, each with 20 numbered days, plus one 5-day month named *Wayeb'* (365 days total). As you might expect, days are also animated and influence daily life. *Wayeb'* is a precarious time because it is the transition between death and life, or the liminal period between the end and beginning before renewal. The *Tzolk'in* and *Haab* mesh once every 52 years (i.e., line up with the same sequence of numbers and days), which is called the Calendar Round. The Maya also keep track of the moon, Venus, and constellations. Those who keep track of the calendar are called *aj q'ij* in K'iche', translated as "he/she of days," or day keepers.[15]

The ancestral Maya engaged with ancestors, gods, and fellow guardians via ceremonies. The sheer number of artifacts deposited as animating caches and de-animating deposits in buildings, funerary offerings in burials, as well as offerings in caves, *cenotes* and other water bodies, *milpas*, and forests, is astounding. The ritual depositing of artifacts impacted production, consumption, and distribution patterns of jade, ceramics, chert, obsidian, shell, and perishable items, including feathers, cotton textiles, botanical medicines, spices, cacao, *copal*, and food. At the ancient Maya city of Dos Pilas in Guatemala, for example, archaeologists led by Arthur Demarest found that 20 to 50% of the artifacts came from nearby caves, including one-third of all ceramics and more than one-half of jade objects. And this tally includes artifacts from temples, palaces, tombs, and houses. Yet the Maya still did not overuse resources.

Since most of the ceramics used in ceremonies and deposited in caches came from the domestic sphere, such as jars, plates, or dishes, the Maya needed to replace them. But before the Maya used them in ceremonies, they first had to de-animate them as household items to release their souls with, for example, a "kill-hole" in the center of a vessel base, and then animate them as offerings. In other words, their life history changed, and it must be acknowledged. If the Maya manufactured objects specifically as items to be cached, then there was no need to de-animate them, only to animate or dedicate them through burning incense and thanking gods and ancestors. Everything has a life history punctuated by rites of passage. The Maya de-animated or released an object's *ch'ulel*, as seen with the Maize God dish, indicated by the kill-hole in its center (see

Figure 4.2). People, objects, fields, monumental and small buildings, pilgrimage destinations, plazas, and cities were de-animated before they were renewed via funerary rites, kill-holes, breaking of items, tearing down of buildings, or burning. Destruction begets creation. Death begets life.

Maya technology did not change much for millennia—for agriculture, ceramic manufacture, and stone tool production—which helped prevent the overuse of resources/fellow entities for millennia. For instance, stone tool materials and types did not really change much over the millennia. Chert was always their main source of stone to manufacture implements, outcrops of which were readily available throughout the karstic landscape. There was a major addition to the stone tool assemblage after 900 CE during the Postclassic—arrow points. Perhaps this had to do with postcollapse, post–urban diaspora life, such as a greater reliance on smaller game and/or other hunting weapons such as blowguns (and clay pellets) and slingshots. While ceramic manufacturing techniques largely remained the same, styles did not (e.g., different clays and tempers, slips, designs, and vessel forms). They changed frequently enough that we can use them to date archaeological contexts within 50 to 100 years or so.

One of the most continuous interactions the Maya have is via openings in the Earth. These portals have some of the earliest evidence for ceremonies, dating to at least 1200 BCE. This is not surprising given that much of the Maya origin history in the *Popol Vuh* takes place in the Underworld. The complex ways with which the Maya engage portals highlight how everything not only intersects, but also has varied aspects depending on the context or life events, as I show with how and why the ancestral Maya killed and deposited witches in portals.

Portals to the Underworld

For Maruch Vet, Sold to a Cave:
Mother of the Night,

Father of the Night,
Great Star of Venus,
Mother Month,
Mother Moon:
Get up, Kajval [Earth God]!
Put on your best clothes.
Let Maruch Vet's body
out of where she's scared to death,
sold to a cave,
sold to a mountain.
Ok, Kajval,
our word, our prayer
is here before you
inside your cave.
We offer you these feathers,
this dove in exchange for Maruch Vet, Kajval
A little food for you, a little drink,
Guardian of the Sacred Cave:
Keeper of the Holy Mountain:
Let go of her body.
It's tied up, shut in,
getting skinnier every day inside the hill, Kajval.
This burden of hers
isn't getting any lighter, Kajval.

ANTEL PÉRES OK'IL[16]

The ceiba is the tree of life. It connects the three realms; its roots reach deep through the Earth into the Underworld. Its trunk and branches burst through the Earth toward the Upperworld. The Maya engage with the Underworld through openings in the Earth or *ch'e'n* (portals), particularly caves and water bodies. *Ch'e'n* are animated, whether they are in villages, towns, ancient cities, or the jungle. The millennia of seasonal rains have created thousands of such portals in the porous limestone bedrock. Some demand pilgrimages, either yearly or for specific reasons, such as during severe droughts.

Ch'e'n also ritually define borders and manifest community identities. Ancestors who reside in lineage mountains are accessed through caves. Gods dwell within them where they control lightning, rain, clouds, wind, the land, and fertility. Chahk, for instance, lives in such openings. Caves are sources of water—streams, springs, rivers, pools, or waterfalls. Mists, rains, and clouds also originate from *wits* and *ch'e'n*. They are places of creation from where humans emerged, as did maize. The sun sets or dies and rises or is reborn through portals. Processions up to the summit of *wits*, whether forest hills or urban pyramid temples served the same purpose—to reach the denizens of *ch'e'n*.

The Maya built pyramid temples and other urban buildings over *ch'e'n*—as an *axis mundi*. Some portals are powerful and must be approached carefully. Maya pray to gods and ancestors at portals, asking them to bring plentiful rain, bountiful crops, and health. In the past, they left offerings, especially jars, some purposefully broken, some with food (e.g., maize), and others with water. The Maya associate jars with Chahk and Ixchel, the Moon Goddess, and their role in the creation of the world. Also, their physical features are like caves (on their side) and *cenotes* (upright).

Ch'e'n continue to be places of pilgrimage for traditional Maya. Worldwide, people from all walks of life often take a long and arduous journey to pilgrimage destinations. For the Maya, they interact and engage with divine and ancestral entities to ensure the continuance of the world.[17] The Maya intensified their visits to some *ch'e'n* in the latter part of the Late Classic period (c. 700–800 CE) and during the Terminal Classic (c. 800–900 CE) in response to the several multiyear droughts that struck the Maya area, as I show in the next chapter on the pilgrimage destination of Cara Blanca, Belize, and its watery portals.

While Maya *ch'e'n* are considered sacred, they also have dangerous qualities because of what can take place in them, such as the extraction of souls through witchcraft, or what they house, such as disease-producing forces. Evil and dangerous spirits also inhabit caves in isolated and unprotected areas—that is, in dark zones. Food is placed at cave entrances as offerings to the spirits so that they can carry illness into the cave and subsequently into the Underworld. Clearly, the ambiguity of caves makes sense given the multidimensional and enmeshed aspects of the Maya inclusive worldview, such as good/evil and dark/light.

The Maya also placed the dead in *ch'e'n*. Death itself is a multidimensional and complex state of being. Take witches, for example. When imbalances occurred, as they invariably did, balance had to be restored, even if doing so required violence against fellow community members.

Maya Witches

Thou shalt not suffer a witch to live
                                            Exodus 22:18, Christian Bible

In 2020 Domingo Choc Che, a Q'eqchi Maya and an *ajilonel*, or a specialist in Maya medicine, who was collaborating with a pharmaceutical project at University College London, was tortured and burned alive in Chimay, Guatemala, because Maya Christians in his village believed him to be a witch.[18] Different worldviews collide. But witchcraft persecution existed long before the Spanish arrived.

Witchcraft persecution occurs globally, and probably has since the dawn of humanity. Witchcraft persecution is when a community blames someone when misfortune occurs—bad weather, failed crops, or famine.[19] Sometimes people blame the gods. If gods are to blame, people pray, proffer offerings, and perform ceremonies. What about when people blame other people? When people accuse someone of causing misfortune, that is, of being a witch, they must pay to restore order or to bring things back to normal—through bullying, ostracization, sanctions, gossip campaigns, and the ultimate price, violent death. Because witches are defined differently than the rest of society, they are treated differently after death as well; for example, in

Europe, people cannot bury accused and killed witches in consecrated ground (church cemeteries).

Historically, witchcraft persecution intensifies during periods of climate instability, as was the case in Medieval Europe and during the infamous Salem witch trials in New England during the late seventeenth century during the Little Ice Age (1300–1800 CE).[20] And this persecution continues today, as illustrated by the Chimay example just mentioned.

When does community consensus turn a mother, father, husband, wife, sister, brother, daughter, or son into a witch? In Amatenango, Chiapas, Mexico, anthropologist June Nash recorded 36 murders during her fieldwork among the Tzeltal Maya in the 1950s and 1960s, over half of which had been accused of witchcraft, many of whom were men.[21]

Traditionally, the Maya view ritual specialists or healers with ambiguity since they perform curing rites in caves—as far as they know, anyway. Anyone with any degree of specialized skill or power over others can be suspect, including political leaders, shamans, and mediators with the Earth Gods. It is particularly dangerous to visit dark zones of caves, so ritual specialists rarely transgress into such areas, but stick to areas with natural light.

Witchcraft takes place in caves. Somebody with evil intent can perform a ritual inside a cave to "sell" someone's soul to the Earth Lord.[22] For the Tzotzil Maya of Chiapas, witches perform rituals in caves using candles, rum, and incense. In highland Guatemala, witches perform the *mal entierro* (evil burial) ceremony in caves with the goal to kill perceived enemies.

Ritual violence often is the response. And it is not enough to kill witches. They also must be kept away from the living and contained in openings in the Earth. For example, in 1996 in Chiapas, Mexico, a mob hanged a man because they believed he was stealing peoples' souls and hiding them in bottles in a cave. At the height of the paramilitary killings in Chiapas, Mexico, in the 1990s by *cortacabezas*, or head cutters, two young men from the village of San Pedro Chenalhó were stoned to death for bringing discord to the community by bragging about being Zapatista supporters.[23] Their bodies were unceremoniously dumped in a crevice—another type of opening in the Earth. In 1648, people witnessed a battle between animal spirit companions—a puma and jaguar—of two Pokomchi Maya of Guatemala. One died of his wounds and the other eventually was hanged for the "murder," and his body was placed in a shallow grave in a ditch rather than in consecrated ground in the church cemetery.

Typically, the dead play vital roles in the daily lives of the Maya as ancestors. Only select family members were buried in the home as ancestors, including men, women, and children. Others were buried in shrines, tombs in pyramid temples, *chultunob* (dry storage features excavated into the limestone bedrock in settlement areas), caves, and other places we have yet to find. Those who were buried near the living helped to maintain social mores and establish a sense of place and community. Those who were buried deep inside portals did not.

The Maya deposit animated objects imbued with sacred/dangerous qualities away from the living. For instance, the Lacandon Maya in Chiapas, Mexico, dispose of god incense pots in caves because they still have dangerous animated qualities due to their role in transmitting offerings to the gods, even though they have been de-animated.[24] Killed witches are treated similarly, apart from their not being de-animated (i.e., no funerary rites).

Witches differ from sacrificial victims: the latter were innocent. In the rare instances when the nonroyal Maya sacrificed someone to the gods, this person was a valuable community member who went straight to the Upperworld.[25] In sixteenth century colonial Yucatán in Mexico, Bishop de Landa noted that Maya leaders sacrificed their sons when the rains were late or during a famine by throwing them into *cenotes*.

Maya witches, in contrast, were considered guilty of causing whatever troubles their community faced. And it was not enough to kill them. Like dangerous animated entities, they were kept safely away from the living and left in portals. Yet both witches and sacrificial victims were violently killed (e.g., heart extraction, strangulation, and head trauma). So how do we distinguish witches from sacrificial victims (and ancestors) in the archaeological record, especially in caves? We do so by assessing how the Maya treated them after death—whether they were buried or not, whether they were placed in light or dark areas, whether the Maya placed funerary offerings with them, or whether they were positioned or haphazardly placed. The Maya treated the remains of those sacrificed reverently and buried them either in the built environment (e.g., house, palace, temple, ball court, or plaza) or in portals. Ancestors and sacrificed persons were not stripped of their identity per se, while witches were—or at minimum were redefined, as I illustrate.

Actun Tunichil Muknal is a cave in central Belize in which the Maya left offerings, especially between c. 700 and 900 CE. It goes 5 km (3+ miles) deep into a *wits* with a stream flowing through its main passage. Most of the offerings are found in the main chamber (c. 200 × 50 m, or 656 × 164 ft) and its smaller adjoining rooms and alcoves. Archaeologists found individuals of both sexes ranging in age from one year to middle-age. And they all were deposited in dark zones.

Only 3 of the 14 individuals were "properly" buried, that is, positioned (e.g., extended, face-up) with funerary objects and entombed (covered by dirt, placed in a tomb or in a cist grave). The Maya hid some of the deceased on ledges or in alcoves, or left others out in the open on cave floors. Most show evidence of trauma, especially on the crania. For example, one of the adults, a female with cranial deformation—a sign of elite status—was killed, likely with a blow to the head, and left sprawled out, face-up, on a ledge that's difficult to reach and not visible from below. Unfortunately, since her remains are encased in calcite and cemented to the surface, it is not possible to determine her exact cause of death. The Maya placed her hurriedly either by throwing or pushing her. She was not accorded the usual rites of a sacrificed person or an esteemed ancestor to-be, especially given her presumed high status. Her presence is even more interesting considering that women today are prohibited from entering caves—during life, anyway.

We find a different story at Actun Uayazba Kab, where the Maya carefully placed and buried the deceased along with funerary objects. This cave is located c. 700 m (0.43 miles) from Actun Tunichil Muknal along a ridge close to the small settlement of Cahal Witz Na and dates to the same period. It is more of a rock shelter with lots of natural light and elaborate petroglyphs, carved anthropomorphic masks, painted triangles, and negatively painted handprints. Archaeologists excavated seven burials with individuals of both sexes ranging in age from an infant to a 40-year-old. The Maya buried all but one underneath a plaster floor along with funerary objects in areas with natural light. None of the deceased have evidence of trauma—that is, they died a non-violent death.

Actun Tunichil Muknal contains witches. Actun Uayazba Kab contains ancestors.

The postmortem treatment of witches reflects their redefined role in life—someone who brought misfortune according to the consensus of their peers. Disposing of witches in dark areas was necessary to contain their perceived evil or wicked forces because they posed a danger to the living. Actun Tunichil Muknal and other caves go deep into *wits*, closer to the dark and dangerous realm. Witchcraft persecution is defined as the punishment of someone in the hopes that things will change back to "normal." By killing witches, the Maya deactivated their perceived evil power by ritually killing and disposing of them in the same place they were believed to have performed the malevolent ceremonies that brought ill fortune. They were never given proper funeral or de-animation rites. Their former family and neighbors left them exposed to dangerous forces, stripping them of their identity. Their souls are trapped forever in this netherworld, never to be recycled or reborn, caught between the worlds of the living and the dead. People condemned former loved ones or neighbors who were then killed as witches to be slated to serve the malevolent gods of the Underworld for eternity.

How do witches fit into the Maya inclusive worldview? One of the hallmarks of this worldview is its emphasis on maintaining a balance to ensure the continuity of the Maya and their nonhuman brethren. Like gods, ancestors, and forest spirits, witchcraft persecution keeps people in check. The message is clear: stay on the "righteous" path or pay the ultimate price, forever. Simultaneously, witchcraft persecution also occurred under circumstances when people blamed others in their community for some mishap such as failed crops or illness or death in a family. Like other forms of conflict globally, accusations of witchcraft are expected to increase due to the detrimental impacts of climate change and decreasing resources.[26]

The Maya inclusive worldview embodies diverse interactions and overlapping and interwoven relations, all for the same purpose—world maintenance. The moral of the story: it takes everyone fulfilling their duties to ensure the continuance of all entities. Compliance is required for the benefit of everyone. Noncompliance is dealt with by violent means. This compliance takes many forms, including reciprocal relations the Maya maintain with the three realms, especially the forest via collaboration and pilgrimage.

Chapter 5

Relations with the Three Realms

For centuries, jungle explorers and sightseers have waxed poetic about the exquisitely plumed birds and the plethora of flowers, butterflies, trees, and bushes. They oohed and aahed about jungle life—colorful birds, coatimundi, howler and spider monkeys, deer, tapir, peccary, and jaguars. They squirmed when they saw tarantulas, scorpions, snakes, flying cockroaches, and crocodiles. And, of course, there were seemingly millions of buzzing flies, mosquitoes, and other crawling and flying creatures.

The Maya see more—so much more—comparable to a pharmacy, a food market, a hardware store, and a general store. The Maya landscape is not a map of resource extraction zones, but a vibrant place pulsating with life. No metals are found in the karstic Maya landscape. Nor were there beasts of burden, and thus, there were no plows or carts before the invasion of the Spanish. The Maya relied on human labor and stone tools and a wealth of knowledge compiled from millennia of their forebears' knowledge.

The ancestral Maya had diverse direct and indirect relations or interactions with the Upperworld, the Earth, and the Underworld. Direct interaction occurred via an enduring collaboration with forests that involved culling and fostering certain flora and fauna, clearing land, fishing, extracting resources, hunting, controlled burning, and gathering construction materials and wood for fuel. Indirect interaction occurred via pilgrimage, as I illustrate with Cara Blanca and its plethora of watery portals in central Belize. Throughout the world, people are prohibited from building, hunting, collecting flora, or planting crops in such sacred places. Consequently, flora and fauna flourish. Thus, pilgrimage destinations *are* conservation areas. Both types of interactions serve the same purpose—the maintenance of the Maya, the tropical forest and its inhabitants, and the world.

The Tropical Forest

Tropical forests epitomize high biodiversity with their abundance of fauna and flora, with which the Maya had a myriad of relations. For example, when the Yucatec Maya talk about the uses of nearly 1,000 plant species, they "often included the uses made of plants by woodpeckers, warblers, or other animals."[1] Often flora have several uses, such as is the negrito (*Simaruba glauca*) tree with its edible and tasty purple grape-sized fruits. The Maya also use its seeds to make soap. When Juan Antonio was explaining its uses, as well as pointing out other interesting trees and undergrowth, a field school student said with amazement, "the jungle is like a refrigerator." Indeed. Palm trees like cohune or Corozal (*Attalea cohune*) provide edible heart of palm and palm oil, as well as leaves for shade, waterproof thatch roofs, and brooms. They also have medicinal properties and produce edible nuts that also can be carved into *objets d'art*. The Maya made diverse uses of limestone—to make stone blocks, mortar, and plaster for floors and walls. They soak and cook maize with limestone to soften it for grinding and to add essential nutrients (calcium, niacin or vitamin B, the amino acid lysine) and to reduce mycotoxins produced from mold.[2] Maya potters used crushed limestone as a temper (additive for stabilizing pottery during firing) when manufacturing ceramics.

The allspice (*Pimenta dioica*) tree has multiple uses as well. Its dried berries are ground and used as an aromatic spice and as a medicine for gastrointestinal ailments. Its leaves are aromatic as well—and they repel insects. The ancestral Maya used allspice berries in their ceremonial cacao beverage, and their leaves were used as designs on royal textiles. Today, 30% of the world's allspice comes from Central America.

There are lots of subin trees (*Acacia hindsii*). Tea made from boiling its bark is used to treat stomachaches and hangovers and to slow the spread of fer-de-lance venom. It has a dangerous side as well. As mentioned in Chapter 1, subin trees are covered with hollow thorns in which ants (*Pseudomyrmex ferrugineus*) reside. You only see the ants when they come out and bite anything that touches any part of the tree—and release formic acid when they do so.

While first aid kits are a must in the field, they sometimes are unnecessary due to the surrounding jungle pharmacy. For instance, the leaves from a low-lying bush (*Piper aducum*), when chewed to a pulp, can be used as a poultice on infected bites, lessening the itching immediately and healing bites by the next day. Drinking tea made from the bark of the Billyweb (*Sweetia panamensis*) tree treats various kinds of fevers, something I learned from Narciso Torres, a Yucatec Maya from Santa Familia in west-central Belize.

The Maya traditionally use soldier ant pincers as sutures. Soldier ants of any species are quite aggressive in protecting worker ants while they are foraging and consequently have disproportionally large pincers that automatically snap closed upon contact with, well, anything. The Maya would capture these warriors and place them, one by one, on an open wound. When their pincers bit into the flesh on either side of the wound, they pulled, removing the body of the ant while leaving the head with its pincers closed in a death grip. Pincers are sharp, clean, and make perfect stitches.

There are all kinds of treatments, not necessarily medicinal. According to Cleofo, Maya put sprigs of the *dormirlona* (*Mimosa pudica*), or the "go to sleep" plant, on a baby's pillow to help them sleep. Why? Because the leaves of the plant fold in on themselves when touched and open a few minutes later—as if they are going to sleep and reawakening.

There is a plethora of edible delights in the jungle—fruits and nuts, stalks and roots, fish, and game. Since the Maya today use over 500 native plants from gardens,[3] fields, and forests, I only mention a few to convey the diversity of foodstuffs. For fruits and nuts, for instance, there are cashew, soursop, coconut flesh and juice—the latter plain or with rum—apple bananas, craboo or nance (*Byrsonima crassifolia*), *anona* (custard apples), mangos, papaya, pineapple, mammee apple, cacao pulp (a tasty white gelatinous goo that encases the seeds inside the pods), and *caimitos* (star apples with their pink-red skin and white interior that taste like white grapes). Jicama, picaya, palm hearts, and wild sugar cane are a few examples of stalks and roots. And there are all kinds of fish from rivers and forest game, including armadillo (*Dasypus novemcinctus*), venison (*Odocoileus virginianus*), gibnut (*Dasyprocta punctata* or *Cuniculus [Agouti] paca*), *tepecintl* or peccary (*Tayassu tajacu*), and iguana (*Iguana iguana*). These examples demonstrate the nutritional and caloric variety conducive for a healthy lifestyle.

The forest provides an array of other items for daily use—natural fibers for mats, clothing, rope, twine, and hammocks. Take *kapok*, from the ceiba tree (*Ceiba pentandra*). It is the billowing fiber that protects seeds that the Maya use to make pillows—and is buoyant enough to use to make life vests. It truly is the "tree of life." Alas, archaeologists won't find any evidence of such items in the archaeological record. The humid tropics are not kind to organic materials; they decompose quickly.

There also are hardwoods for posts and vines to latch them, clay to make ceramics, minerals and organics for paints, chert outcrops the Maya quarried to make stone tools (e.g., drills, hoes, axes, and spear points), and tree resins either burned as incense in ceremonies or collected to make the rubber balls used in ball games in the ball courts found in every Maya city. Throughout the forest, one comes upon large X-shaped scars on rubber (*Castilla elastica*) and *chicle* or sapodilla (e.g., *Manilkara zapota*) trees, where Maya tapped them for latex and gum. Resin from the *copal* tree (*Protium copal*) is still used as an incense (*pom*) in nearly every ceremony; when lit, its fragrance is wonderful—a potpourri of jungle spices.

There are many other ways the Maya utilize their extensive knowledge. For instance, Antonio, a Maya who worked at Banana Bank, told me that if you hang a dead coatimundi (*Nasua narica*) in *milpas*, it keeps other coatimundi away because they do not like the smell of their own dead.

***

After the Maya abandoned cities and kings by 900 CE, the forests, with relatively few Maya about, recovered and expanded into deserted cities and farmsteads that gave rise to the forests we see today. In Chapter 1, I mentioned Colleen and Cleofo's botanical collection project. It is relevant mentioning it again here. They collected over 300 botanical species not only from areas with and without archaeological sites between Yalbac and Cara Blanca, but also from Valley of Peace Village home gardens to assess how the forest composition reflects millennia of interaction. Cleofo identified 95% of the botanical specimens as fuel, medicines, construction materials, edible plants, nuts, fruits, and more. They found more "useful" flora at sites than in areas without sites, which makes sense given that the Maya collaborated with the forest and promoted certain species. Flora without a use do not have names, but they are still respected. In fact, there is no Mayan word for "weed," just an "un-useful" plant.

Their results show that the forest today indeed reflects ancient forest collaboration. For instance, the species most frequently found near ancient houses include those used in ceremonies (e.g., *copal* tree) and for food—for example, picaya (*Chamaedorea* spp.), ramón trees (*Brosimum alicastrum*), wild plum (*Spondias* spp.), and mammee (*Pouteria mammee*). They found a variety of species that would have been used in the house, for example, to make brooms, ties, and thatch, including bayleaf palm (*Sabal* spp.), cohune palm (*Attalea cohune*), broom tree (*Cryosophila stauracantha*), and vines of all sorts, including the water vine, which literally is filled with water (*Vitis tiliifolia*). And, not unexpectedly, they collected specimens that had medicinal uses, such as mushrooms (*Ganoderma lucidum*).

Archaeologists often find ramón trees near Maya sites. Their shade would have provided a respite from the hot sun. They produce edible fruit, and its seed, referred to as a ramón nut or Maya nut or breadnut, is gluten-free and has a high content of vitamins A, B, C, and E, calcium, fiber, potassium, iron, zinc, folate, protein, and antioxidants. The Maya dry the seeds in the sun, after which they can store them for up to five years. When cooked, the nut tastes like mashed potatoes. When roasted, it tastes like coffee or chocolate. Ground ramón nuts are used to make gruel and tortillas. It is an ideal food source during long periods of drought or other times of food insecurity.

There were other patterns as well. For example, *chechem*, or poisonwood (*Metopium brownei*), is usually found in uninhabited areas, but rarely near sites. However, gumbolimbo trees (*Bursera simaruba*), which typically grow near *chechem* trees, are found near sites (also known as the "tourist" tree because its bark is red and peels like tourists who spend too much time in the sun and get burned). The resin of the former is the antidote to the caustic resin of the latter, the wood of which is a hardwood with a satin sheen and beautiful iridescent color. *Chechem* poisoning can result in a painful rash, blisters, and respiratory problems. The Maya encouraged the growth of gumbolimbo trees, the resin of which also can be used to treat other ailments.

The reciprocal relations the Maya had with the tropical forest endured because they did not misuse forest entities. The Maya maintain the forest, and the forest maintains the Maya. One means is via pilgrimage to portals, including watery ones like Cara Blanca with its 25 water bodies. Cara Blanca is a fascinating place for other reasons as well, including its high biodiversity, year-round water availability, nearby fertile soils, and how the ancestral Maya interacted with it. In fact, Cara Blanca's high biodiversity is a direct result of this type of interaction, as I explain.

Pilgrimage to Watery *Ch'e'n*

Water is life. It is vital. It is vibrant. Water in *cenotes* emerges from the Earth. The names of Classic Maya cities often include references to watery places or mountains. Maya collect *suhuy ha'* (pure water) from *cenotes* or caves for curing ailments and for rain ceremonies. In fact, "waterholes are highly sacred, and myths are told about each of them, describing the circumstances under which the ancestors found the water and the ways in which the waterhole acquired its distinctive name."[4] In Zinacantán, Chiapas, Mexico, the Tzotzil Maya mix water from their seven sacred water bodies for large ceremonies, and from one or two for curing ceremonies. Everything is about renewal. Thus, watery places demand pilgrimages. Pilgrimage is when people from all walks of life take an often arduous journey to special places to visit and engage with the gods or ancestors or other "supernatural" entities for either renewal or maintaining balance in the world.

The Maya take a scripted journey, in this case, a ceremonial circuit.[5] A ceremonial circuit is a procession that has stops along the way, as found among the Yucatec Maya of the Yucatán, where the circuit mirrors the path of the sun, east to west and counterclockwise. Each stop along the way requires its own ceremony and offerings. The journey connects people and places and renews relations with the tropical forest and its denizens. Processions to *wits* (lineage mountains) and *ch'e'n* (portals) underscore ancestral land and water rights and obligations (i.e., relations). Urban *sak b'eh*, or causeways, served the same purpose in Maya cities. Religious and community leaders undertake such circuits as part of major ceremonies (e.g., New Year, End of Year) and under certain circumstances, such as severe drought.

The Sacred Cenote of Chichén Itzá in the northern lowlands is a well-known pilgrimage destination. It was dredged in the early twentieth century, a practice that would not be allowed today due to the extensive damage it causes. Archaeologists extracted ceramic figurines, copper bells, jade, Postclassic gold and silver objects from Mexico, *copal* incense balls, wood items, shell, textiles, chert and obsidian, rubber, and at least 228 ancestral remains representing centuries of offerings up through the Spanish invasion, including people from throughout the Maya area, the Gulf Coast, and perhaps even the central Highlands of Mexico.[6]

The Maya left offerings in lakes as well, as they did for 2,000 years at Lake Amatitlán in Guatemala, where divers found over 400 ceramic vessels, many dating to the Classic period (c. 300–900 CE), with designs portraying various flora and fauna including snakes, flowers, lizards, fruits, and spider monkeys. Ceramic vessels also depicted human heads, Chahk the Rain God, Tlaloc the central Mexican Storm God, as well as fertility and death gods. Another fascinating aspect of Lake Amatitlán is that people from different ethnic groups from throughout Mesoamerica came here—from the Maya area, Teotihuacan in central Mexico, the central Mexican highlands, and other regions.

There are many such examples, including Cara Blanca ("white face"), where the ancestral Maya left a minimal human footprint—even though water and nearby agricultural soils were plentiful. There also would have been an abundance of flora and fauna that would have supported many people—fish (e.g., cichlids), turtles, frogs, toads, snakes, snails, mollusks, crabs, crocodiles, waterfowl (e.g., egrets, cormorants, and herons), all kinds of birds (e.g., toucans, hawks, trogons, and parrots), deer, peccaries and other game, jaguars, and cattails and other aquatic flora. And water lilies abound on pool edges, indicating clean water.

The vibrant landscape of Cara Blanca encompasses 25 lakes (2–17 m, or 7–56, ft deep) and *cenotes* (15–62+ m, or 49–200+, ft deep) that run east-west along a fault at the base of an escarpment c. 100 m high in parts (328 ft) (Figure 5.1). The *cenotes* are steep-sided

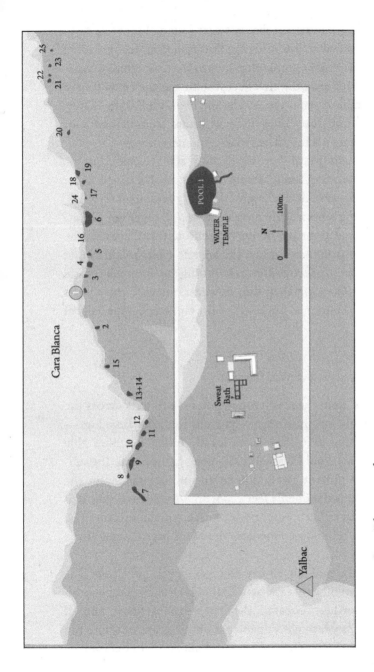

FIGURE 5.1 *Cara Blanca pools*

For scale, the size of Pool 1 is 100 × 70 m or 328 × 230 ft. Today much of the area south of the pools has been deforested (see Figure 1.2). Yalbac is on the lower left. The inset image includes the sweat bath, its associated settlement, and Pool 1 ceremonial buildings.

Source: Illustration by Julie McMahon. Courtesy of VOPA.

sinkholes fed by groundwater that last year-round. They were fashioned by the gods with additional modification from the impact of the Cretaceous-Tertiary asteroid that created the Chicxulub crater—the one that started the domino effect resulting in the extinction of dinosaurs 65 million years ago. While the *cenotes* are immediately surrounded by soils too clayey to grow crops, well-drained soils great for agriculture are found less than a kilometer away.

Lots of ancient Maya farmsteads and elite compounds are located near the western lakes, but not near the *cenotes*. *Cenote* water levels barely drop, even at the height of the dry season when creek, lake, and river levels noticeably drop or dry up all together; I know this since our excavations and diving expeditions take place near the end of the dry season in mid-May. The ancestral Maya didn't build cities or houses or plant crops here even during the Terminal Classic droughts (800–900 CE). They did, however, increase their visits here and added ceremonial buildings to the landscape, especially at Pool 1.[7]

The Cara Blanca Ceremonial Circuit

To get to Pool 1 from the south today, you have to walk across the remnants of an old logging bridge—basically just a few large logs—to get to the other side of what is locally called the Blue Nile (Labouring Creek). Pool 1 is about a kilometer to the north. The sudden cool breeze gives it away. Whenever you are walking in the sweltering jungle and suddenly feel a cool breeze wash over you, you know you are near either a water body or cave. As you walk up a rise toward the pool, a building emerges. The Maya built it right at the pool's edge.

The rest of the building slowly emerges as you get closer. The ancestral Maya would have experienced this as well. They came from all over—the northern Yucatán, the Petén in Guatemala, and all parts of Belize. They brought valuables as offerings to throw into the deep watery portal. They brought revered ancestors to dedicate or animate ceremonial buildings. The Maya came to participate in rain ceremonies.

Over the years excavations and diving expeditions have revealed a different and amazing world, especially at Pool 1 (62+ m, or 200+, ft deep), including a massive underwater cave that goes deep into the cliff face, as well as megafauna fossils, trees that have been falling in for thousands of years, diverse aquatic life, and Maya artifacts. The Maya recognized its significance as well, building several ceremonial buildings, including a water temple.

The first inkling that Cara Blanca was a place of pilgrimage was during excavations of Structure 1, which turned out to be a water temple. Underneath the vegetation and accumulated debris (soils, leaves, and roots) was a layer of loose tufa stones, a limestone created underwater, and burned materials, including hundreds of large broken and incomplete jars and serving vessels that covered the roofless temple. What this means is that all their visits, building programs, and ceremonies were to no avail. The Maya ceased their engagement with Cara Blanca and became part of the diaspora out of the southern lowlands by 900 CE. First, though, they had to de-animate the ceremonial buildings. De-animation was done not so much to end their life history, but rather to prepare them for a different path, one that did not involve the Maya.

Getting to Cara Blanca is challenging today, as it would have been over 1,100 years ago. Maya followed the path of the sun east to west or wove north and south mirroring their cyclical existence. The long and arduous journey served two purposes—getting there and engaging with the tropical forest. To come up from the south would have required the Maya to cross Labouring Creek, a perennial creek that merges into the Belize River further east. To the north are steep cliffs interspersed with narrow, steep, and rocky ravines. The Maya knew when they were getting close to a pool because of the cool breeze—its *ik'* or breath or life. *Ik'* is possessed by all entities, including, for example, trees, maize, the Earth, and people. It is comparable to a "life-force" animating its possessors. All the meanings of *ik'*—wind,

FIGURE 5.2 *Water temple on the southwest edge of Pool 1*

The northeast section of the temple has collapsed into the *cenote*.

Source: Drone photo by Tony Rath. Courtesy of VOPA.

for example—are not seen as separate or isolated from one another, in line with the merged aspects of their worldview.

Before the advent of drone photography, such as this image of the water temple (Figure 5.2), Maya foremen and excavation assistants built ladders to lean against tall trees to take overhead shots. Archaeologists who work in the jungle still use ladders since the dense canopy often makes the drones useless. Yalbac Ranch, which until recently owned Cara Blanca,[8] prohibited trespassers, looting, hunting, and felling trees. But the jungle is near impossible to patrol, which is why nearly every large structure has been looted.

At Pool 1 (100 × 70 m, or 328 × 230 ft), the ancestral Maya added seven structures to the landscape, two of which have been excavated, the water temple and a ceremonial platform, both dating to the Terminal Classic (c. 800–900 CE). They had visited in earlier time periods, but not like this. Desperate times called for desperate measures—which in this instance meant intruding into a place previously largely left untouched. The Maya not only intensified their visits during this century-long period of droughts but changed how they engaged with Cara Blanca.

The water temple is a long building (20 × 7.5 m and 3.5 m tall, or 66 × 25 and 11.5 ft tall) that mirrors the *cenote* edge. In fact, it sits so close to the edge that its northeast section, weakened by extensive looting, has collapsed into the water. Unfortunately, looting always increases after archaeologists start working somewhere. The Maya expended much energy and labor constructing cut-stone walls, thick plaster floors (c. 9 cm, or 3.5 inches), thick cobble ballasts or floor supports (c. 10–13 cm, or 3.9–5.1 inches thick), and thick boulder and cobble fills (up to 1.7 m, or 5.6 ft, thick). It has an intricate layout with a passageway, offset rooms, and only one entrance. The passageway wraps around the temple to guide people to a room where we found offerings (e.g., most of a large, inverted jar, upon which its creator had painted someone—perhaps a priest—in a jaguar costume or pelt) and the remains of feasting.

We know it is a water temple because of what we found—but also by what we didn't find—your typical household items. Just like every home today has pots, pans, serving dishes, bowls and tools, so too did the ancestral Maya. Instead of metal vessels and tools, they had ceramic and stone ones. Nor did we find individual or small serving vessels, *metate* and *mano* sets—used for grinding the all-important maize, or *incensarios* (incense burners) that the Maya typically used in every ritual. Everything recovered was incomplete. Gluing efforts in the lab were to no avail—nothing was complete. More about this in a moment.

Ceremonial feasting took place at the temple, indicated by the large jar and dish sherds recovered (orifice diameters up to 55 cm, or 22 inches). Ceramic styles originated in different areas, including the northern lowlands (e.g., jaguar and water symbolism), the Petén in Guatemala, and other regions. Also recovered were 200 faunal bone fragments, mostly charred, such as bird, deer, and other mammals, as well edible freshwater shells (*Pomacea*). Interestingly,

Pomacea are both terrestrial and aquatic, so they have lungs *and* gills. Did the Maya include them in their ceremonies because of their merged qualities?

This is what I think happened. Priests and participants performed rain ceremonies using large jars to collect *suhuy ha'* (pure water), facing east where the sun is reborn each morning. The men prepared food and drink for feasts—*atole* (a maize gruel), *balché* (made from the bark of a *Lonchocarpus violaceus* tree, which is soaked in water and honey and fermented), and blood sacrifice such as turkeys, deer, or *copal* resin shaped into hearts or animals. The gods witnessed the rites reflected on the shimmering portal's surface—as were the palm fronds and vivid-colored flowers decorating every surface. And the sounds of the accompanying drums and flutes would have reverberated for miles.

When gods and ancestors failed to bring the rains, the Maya de-animated the water temple by dismantling its corbel-vaulted roof and placing the vault stones ($1 \times 0.5 \times 0.15$ m, or $3.3 \times 1.6 \times 0.5$ ft) inside the rooms. They then covered the entire temple with tufa and burned something on top—likely some type of vegetation since the building underneath is not burned. The tufa itself is created in mineral-rich water: calcium carbonate precipitates around items such as fallen branches and the like. As a result, the forms are unique, often looking like limestone coral—as seen in Figure 5.3, where you can see the negative imprints of branches long gone. Limestone is plentiful on the surface nearby, in the ravine, and, well, everywhere. Yet the Maya went into the watery portal to collect the tufa they used to de-animate the water temple.

The Maya also performed ceremonies on a long and narrow ceremonial platform, Structure 3 (7.5×3.7 m and 0.8 m tall, or 25×12 and 2.6 ft tall), located c. 22 m (72 ft) southeast from the water temple. It has a small step that is only a meter (3+ ft) from the water's edge. The Maya would have stood on this step to proffer offerings into the watery portal. It was surprising to find nearly 7,000 sherds from the local area, the northern lowlands, the Petén

FIGURE 5.3 *Tufa, a type of limestone created under water*

Source: Courtesy of VOPA.

in Guatemala, and southern Belize during excavations. The Maya had broken, burned, and left most of the ceramic pieces on the plastered surface—yet none of them, once again, represented complete vessels. The Maya de-animated the platform by sealing it with a thick layer (c. 1 meter) of large boulders (60+ cm, or 24+ inches, in diameter).

One of the most fascinating discoveries during excavations of the ceremonial platform was the ancestral remains. But not burials. Caches. Human caches. Let me explain. We exposed three individuals in three different places within the platform fill, a young adult of indeterminate sex in a tightly flexed (fetal) position, a young adult of indeterminate sex tightly flexed and placed on their right side, and an adult male face up with his legs tightly flexed and resting on his chest, indicating that he, as well as the other human caches, had been bundled (literally wrapped into a bundle with cotton cloth or another kind of fiber long since decomposed).

The remains were left *in situ* since there was no reason to remove them. It is important to respect the Maya dead, as we do our own. Some teeth and small samples of bones were collected to export to the United States (with permission from the Belize Institute of Archaeology) for isotopic analysis to determine their place of origin. Strontium isotope analysis revealed that the three human caches, as well as the only ancestral remains recovered at the sweat bath described below, were of the same age range (16–24 years) and came from the surrounding area.[9]

No funerary objects were found with the remains, nor signs of violence, so they are not sacrifices or witches—or ancestors. The Maya kept ancestors close to home by burying certain family members in or near their homes with funerary objects. Human caches, however, serve a different purpose signifying their inclusive worldview; since every entity or being, including people, is on the same plane and serves to maintain the world, then just as the Maya cache vessels and lithics, they also cache humans as offerings. These were revered but deceased personages. Among the Yucatec Maya, bundles are used to communicate with ancestors, gods, and other entities. The Postclassic K'iche' Maya took bundles of gods or ancestors on pilgrimages and when they migrated to new areas. In this case, Maya brought their most valuable possessions as offerings and as a means to communicate with ancestors and gods.

The Maya also de-animated the water jars used in ceremonies. They broke off thumb-size pieces from jar necks and shoulders. The Maya then either took some of the broken pieces with them as a remembrance to their postdiaspora homes, deposited them somewhere else, or threw them into the *cenote*. The point is that broken pieces or sherds are just as significant as the whole vessel, and the same goes for other "incomplete" items.

Midway between Pools 1 and 2, less than a kilometer apart, is an architectural compound that includes multiple rooms and a sweat bath for ritual purification (see Figure 5.1); being in the humid tropics, you would not need one for anything else. The Maya would have added fragrant plants to the water they poured on the hot stones to create steam, thus creating their own mist or clouds from the *ch'e'n* (portal) they built. The sweat bath itself is nearly circular (3.7 × 3.7 m, or 12 × 12 ft) with what remains of a true arch c. 1.8 m (5.9 ft) high. True arches are rare in the Maya area. Most larger buildings have vaulted roofs that rely on counterweight stones for support.

The sweat bath abuts a long structure 40 m (131 ft) in length with at least five rooms and likely more. It is difficult to tell because the Maya de-animated the sweat bath compound by dismantling the roof—as well as many of the walls and floors. This goes beyond the usual building de-animation, but the Maya must have had a good reason to do so. They still covered the entire compound with cobbles and small boulders and topped it off by placing the body of someone local based on strontium isotope analysis. Just west of the sweat bath compound is a group of five low residential mounds and a *plazuela* (open area encircled by structures) that suggest short-term occupation. Visitors probably stayed here, where they would have slept, eaten meals, met friends, made new friends, acquired keepsakes, and ritually prepared for ceremonies at Pool 1 and perhaps other *cenotes* along the ceremonial circuit.

Another stop along the ceremonial circuit included the escarpment above Pools 14 and 15 (both *cenotes*), at the top of which the Maya built a water shrine. The shrine sits high above a deep sinkhole about halfway down to Pool 15 that started out as a cave whose roof has recently collapsed; so, any offerings are deep in the Earth where they belong, inaccessible to us. A crew rappelled over 40 m (131 ft) down to explore the aptly named Motmot Sinkhole (apt because of the resident motmot birds [*Momotus momota*]), where they found a small cave at the bottom that had a pool of water that probably connects to Pool 15 further downslope. Interestingly, a recent PBS show, "Wonders of Mexico: Forests of the Maya," had a brief segment on motmot birds and mentioned that the Maya use their calls to determine the location of *cenotes* since they often nest in the soft limestone sidewalls. Yet another reason the Motmot Sinkhole is aptly named.

In the hills above Pool 6, a lake created from two collapsed *cenotes*, the Maya built a small compound with a plain *stela* (a carved stone monument; any painted designs that may have existed have long since faded) and an altar (a circular stone with a flat surface) that dates to the Terminal Classic period. *Stelae* (plural of *stela*) are typically only found in cities. Another stop on the ceremonial circuit is Pool 20, a *cenote* with a diameter of 100 m (328 ft) and depth of 40 m (131 ft)—and a megafauna fossil–laden clay bed. The Maya transformed a hillside 40 m north of the *cenote* to create a large platform (38 × 26 m, or 125 × 85 ft) upon which they reshaped a natural knoll in the same time frame as other Cara Blanca ceremonial architecture. The Maya only added shaped stones when necessary to get the shape they were after—a pyramid temple (22 × 12 m and 3.5 m tall, or 72 × 39 × 11.5 ft) with a staircase carved into the limestone bedrock. They merged the built and "natural" to transform them into something else—something that served a purpose along the ceremonial circuit. There are other stops as part of the ceremonial circuit that have yet to be explored.

The Maya did not leave a noticeable footprint at Cara Blanca *cenotes*. What about under water?

Diving into *Ch'e'n*

With Pool 1 being so massive and deep, it was difficult deciding where the divers should explore for Maya offerings in 2010, especially prior to knowing about the ceremonial platform step exposed in 2016. With its depth, tangle of trees, changing visibility, and crocodiles meant that divers had to be extra careful. But this is what exploration is all about. In addition, divers also used a hydrolab to determine the mineral and chemical content of the pools; Pool 1 has the freshest water of the eight tested, though it has a high content of minerals.

Divers recovered most of the artifacts—a few jar sherds—immediately below the water temple, likely from looters' debris and building collapse. Divers also discovered another portal invisible from the surface—a massive underwater *ch'e'n* that we named Actun Ek Nen (Black Mirror Cave) on the north wall that goes deep into *wits* (lineage mountain), at least 70 m (230 ft) (Figure 5.4). It is the largest freshwater cave on record in Belize. The top of Actun Ek Nen is over 30 m (100 ft) below the surface, with an entrance 40 m (131+ ft) wide that continues all the way to the *cenote* floor. Interestingly, in 1588 Antonio de Ciudad Real, when writing about the Sacred Cenote at Chichén Itzá, noted that "they even say that in the wall of this well or *zonote* [*cenote*] there is a cave which enters a considerable distance within <the cliff>."[10] Did the Maya know about Actun Ek Nen? After all, deep-sea pearl divers can free dive as deep as 40 m (131 ft). I am sure they did.

The underwater forest tells its own story. Trees have been falling into Pool 1 from the cliffs above for millennia. Thanks to a National Science Foundation grant in 2014, scientists and divers were able to explore this forest. Dendro-climatologist Brendan Buckley (Lamont-Doherty Earth Observatory, Columbia University), with assistance from underwater archaeologist Andrew Kinkella, cave exploration diver Chip Petersen, underwater videographer Marty O'Farrell, and underwater photographer Tony Rath,[11] collected tree specimens, which Buckley has identified as broadleaf species, mostly from the Meliaceae family (e.g., mahogany). Cleofo helped Brendan identify the tree species from the *cenote*, as well as the trees they sampled nearby that were downed in the 2010 hurricane. Divers also collected a piece of wood embedded in a fossil bed that is nearly 9,000 years old based on radiocarbon dating, possibly a broadleaf species similar those found near Pool 1 today. And there are hundreds of embedded fossils in the walls of Pool 1 (and Pool 20) that await future explorations. So far, the only specimens collected are giant sloth species (*Paramylodon harlani* and *Eremotherium laurillardi*), according to paleontologist and diver H. Gregory McDonald.

FIGURE 5.4 *Chip Petersen and Marty O'Farrell about to enter Actun Ek Nen*

Source: Photo by Tony Rath. Courtesy of VOPA.

Explorations at Cara Blanca have just begun. In 2013 divers shared their impressions of Pool 1. Chip Petersen stated that "the *cenote* seems like a living creature; survey lines from previous seasons are mostly gone. Our impact from previous seasons is being erased annually, which presents new challenges each year." Alas, the number of new all-weather roads in this area is concerning; the nearby sugar cane operation and industrial agriculture both rely on extensive land clearing and an array of chemical fertilizers, herbicides, and pesticides.

***

Cara Blanca encompasses land, water, *wits*, sky, and forest. For a brief moment in its life history, the Maya became part of the history of this amazing place. They built a substantial water temple facing east toward the rising sun, as well as other ceremonial structures. The Maya intensified their engagement with Cara Blanca between 800 and 900 CE because they felt it was the only way to address their dire circumstances. And despite the several prolonged droughts, they never built cities or homes or planted crops there. They could have done, which would have allowed them to stay in the area, especially since there are fertile agricultural soils near the pools. But they did not. After the urban diaspora, the relatively few Maya who remained in the area lived in small communities near more reliable water sources—near lakes, for instance. They still did not live near *cenotes*. That is the power of place.

The reciprocal relations the Maya had at Cara Blanca are just as critical as direct interaction with the tropical forest (culling, hunting, collecting, burning, and clearing). The minimal engagement allowed flora and fauna to flourish, which in turn promoted biodiversity and conservation. Pilgrimage destinations like Cara Blanca serve as a reserve for "resources"—a gene bank. In fact, Colleen and Cleofo recovered 37 different plant species near Pool 1, some only found there, including wild papaya (*Carica papaya* L.), jackass bitters (*Neurolaena lobata* [Linnaeus] R. Brown), the bullet tree

(*Bucida buceras* L.), and wild pineapple (*Bromelia pinguin* L.). The Maya use jackass bitters to treat malaria and insect bites, while wild pineapple's fruit is edible and the entire plant can be soaked and beaten, and the resulting fiber can be used as thread. They ridded themselves of intestinal parasites by drinking tea from boiled leaves of the jackass bitter tree—as well as the boiled bark of the copal tree (*Protium copal*)—or by drinking the juice of the Mexican wormseed fruit (*Chenopodium ambrosioides*). None of these species were found in plant surveys near Yalbac or between Yalbac and Cara Blanca. Further, the noticeable presence of medicinal gumbolimbo trees (*Bursera simaruba*) also sets Pool 1 apart. And I am sure the Maya had other practices that promoted conservation: food and hunting taboos and seasonal restrictions, for example.

If you know what you are looking for and know your way around the jungle, you can survive just fine. I think about the Maya doing the same for thousands of years because of their myriad relations with the three realms. The maintenance of the world requires such interactions. They work.

Chapter 6

The Maize People

Thus was found the food that would become the flesh of the newly framed and shaped people. Water was their blood. It became the blood of humanity.... The yellowness of humanity came to be when they were made by they who are called She Who Has Borne Children and He Who Has Begotten Sons, by Sovereign and Quetzal Serpent. Thus their frame and shape were given expression by our first Mother and our first Father. Their flesh was merely yellow ears of maize and white ears of maize. Mere food were the legs and arms of humanity, of our first fathers... and mere food was their flesh.

POPOL VUH, *sixteenth century K'iche' Maya origin history*[1]

The K'iche' Maya origin history, *Popol Vuh*, begins, "THIS IS THE BEGINNING.... We shall begin to tell the ancestral stories of the beginning."[2] "We shall begin" in K'iche' is *tikib'a'*, which translates as "to plant." "K'iche" itself translates as "many trees" or "forests." The preamble continues, using terms like the "sowing [*tz'uk* or germinate/sprout] and the dawning." Newborns also "sprout." As Maya farmers plot out their *milpas* (fields), so too do Earth's creators:

Great is its performance and its account of the completion and germination of all the sky and earth—its four corners and its four sides. All then was measured and staked into four divisions, doubling over and stretching the measuring cords of the womb of the sky and the womb of the earth. Thus were established the four corners, the four sides, as it is said, by the Framer and the Shaper, the Mother and Father of life and all creation, the giver of breath and the giver of heart, they who give birth and give heart to the light everlasting, the child of light born of woman and the son of light born of man, they who are compassionate and wise in all things—all that exists in the sky and on the earth, in the lakes and in the sea.[3]

Maize. It is a living force. It is the stuff of existence. It has its own cycle of life: its seeds are planted, it takes root and emerges from the rich Earth and matures. The image of the ceramic dish in Chapter 4 (see Figure 4.2) depicts this cycle, where the Maize God bursts forth from the Earth, reborn. Everything is connected and is part of the cycle of life—and death.

Maya life revolves around maize (*Zea mays*), as well as beans (*Phaseolus vulgaris*) and squash (*Cucurbita* spp.). They are the "Three Sisters." They complement each other nutritionally (e.g., protein and carbs) for humans and the Earth. Farmers plant their seeds together in the same hole. Growing maize stalks absorb nitrogen from the soils. If nitrogen is not replenished, the soils become less fertile, resulting in lower yields and plants that are more susceptible to diseases. Rhizobia bacteria found at the base of beanstalks, however, replenish some of the lost nitrogen. The squash vines that grow horizontally with their large leaves provide shade to retain soil moisture and prevent extensive weed growth. The Three Sisters are like a well-composed symphony that yields more food than if each of the sisters were grown separately.[4]

This symphony extends to other crops—manioc, tomatoes, pineapple, avocados, tobacco, chili peppers, and many more domesticates (e.g., the protein-rich chaya, *Cnidoscolus aconitifolius*), and nondomesticates. If plants need shade to flourish, for example, the Maya place the seeds or seedlings near a tree. If plants need saturated soils or direct sunlight, the Maya plant them accordingly. Diverse crops (i.e., multicropping) and strategies fed a growing population throughout the Classic period without massive deforestation like we are witnessing at present.

This symphony began 4,000 years ago when the Maya adopted an agricultural way of life. After millennia of collaborating with the tropical forest, they knew it was time to increase their engagement with the Earth to feed growing families and communities. And they knew that the best way to do this was to work with the plants, soil, and seasonal rains, as they had worked and continue to work with the forest in home gardens, *milpas* near and far, and orchards. This sustainable strategy—species-rich tended plots and continued collaboration with the forest as another garden, so to speak—has thrived for millennia.

This is how it works.

Green Farms

The Maya mimic forest biodiversity in their *milpas*, home gardens, and orchards to spread risk. If a disease or drought were to kill off a particular species of plant or tree, there still would be plenty of others that are disease and drought resistant. Today, as I've mentioned, the Maya use over 500 indigenous plants from tended plots and the forest.[5] Over-the-counter multivitamins are not necessary. That said, modernization and the impact of Western influences are changing these practices. From what I have seen over the last few decades, more and more Maya farmers are using chemical fertilizers and pesticides. And, if possible, they hire Spanish Lookout Mennonite farmers to plow their fields, where formerly they left trees for shade and for what else they provide—fruit, nuts, vines, palm leaves, and medicine. Parents focus more and more on education in government schools and universities, and less and less on traditional forest knowledge and farming practices.

What traditional Maya farmers do not rely on is monocropping. Tended Maya plots may appear unplanned and messy based on our structured Western notions of what a garden, field, or orchard should look like. But they are quite organized and reflect a diverse microcosm of efficient interaction, resulting in blurred boundaries between forests and tended plots. Take another look at Cleofo's home garden (see Figure 1.1) to get a sense of what I mean.

Depending on local conditions (slope, thin or thick soils, amount and distribution of fertile soils, and access to water), Maya farmers rely on diverse small-scale extensive (slash-and-burn) and intensive subsistence practices such as raised fields, drained fields, dams, canals, ditches, and terraces in short- and long-fallow fields near and far from home, and combinations thereof. They also farmed some *bajos* (seasonal swamps or wetlands); others are too clayey for agriculture but do provide other resources such as fish, clay, aquatic plants, and bamboo. The ancestral Maya had to walk to their fields since there were no beasts of burden prior to the arrival of Europeans. They also got around using canoes on rivers and larger creeks. That said, rivers are turbulent in the rainy season and quite low in the dry season. Maya walk a lot.

There are a few things to keep in mind when considering Maya farming. The relatively little surface water (rain soaks through the porous bedrock) and topography (e.g., entrenched rivers) are not conducive to extensive and large-scale irrigations systems. And because of the extreme humidity, storing grain is challenging—anything organic quickly rots and decomposes. Maya use maize cribs to store grain and add lime dust to keep the insects at bay since the fine dust suffocates them. Or they smoke the maize to dry it out and store it in *chultunob* (underground storage pits). Some also harvest when they need to by, for example, turning down maize stalks while they are still rooted in the ground to interrupt their food supply, thus drying out maize ears and providing protection from rain and birds and other pests, as well as creating more sunlight, which is critical for bean plants.

We know the Maya used sustainable practices for millennia based on paleobotanical (pollen and phytoliths) and faunal remains (nonhuman bones) in the archaeological record, indicating the presence of healthy forests. Since vegetation is highly sensitive to changes in temperature, sunlight, and precipitation, analyses of paleobotanicals, mostly from sediment cores, allow archaeologists to reconstruct past environmental conditions. Pollen is more inclusive of arboreal vegetation, while phytoliths (silica absorbed from ground water into plants and, eventually, redeposited in the soil) better represent grasses, which either signify dry periods or deforested areas. As the population grew over the millennia, the Maya increasingly intensified their subsistence strategies. Recent lidar mapping has led archaeologists to estimate that up to 11 million people lived in the interior southern lowlands during the Late Classic (c. 600–800 CE).[6] One might assume that the entire area would have been deforested. But it was not. As mentioned, pollen records in sediment cores extracted from lakes, *cenotes*, and *aguadas* (rainfall-fed natural depressions) suggest varying degrees of forest collaboration and deforestation near Maya cities.

After the October 2005 Hurricane Wilma destroyed huge stands of trees in the Yucatán in the northern lowlands, and after wildfires broke out during the dry season several months later, edible plants (e.g., manioc) started to emerge in the newly open areas that had not been settled since the 1600s.[7] Manioc, a tuber, lay dormant for over 400 years. All it took was some sunlight for manioc and other cultigens to bloom once again. Destruction begets creation.

We find the bones of a variety of animal species at Maya sites, such as white-tailed deer, turtles, turkeys, tapir, and peccary, even at the height of population size and density in the Late Classic. Since most nondomesticated animals can only breed and rear their young in forests, their presence in the archaeological record indicates the presence of healthy and biodiverse forests, which is also indicated by long occupation histories like we see at Saturday Creek, where Maya families lived in the same homes for over 800 years (10 years of plowing likely erased another 400–500 years of family histories). Faunal bones (e.g., mammals of all sizes, such as deer, raccoon, gray fox, peccary, rabbits, birds, and coyote) remain diverse in the archaeological record throughout this city's long history.[8] This long occupation would only have been possible if reciprocal relations with local water sources, soils, and forests were maintained—and clearly they were.

The Maya burn *milpas* from about mid-March through mid-April. At the beginning of the dry season, farmers prepare the field by clearing much of the vegetation—certain trees are kept for shade. Once everything is dried out a month or so later, they burn it. Ash is a natural fertilizer because it contains phosphorus, magnesium, and other nutrients. Burning also kills any agricultural pests. They plant about two weeks before the rains are supposed to start. If the rains come too soon, seeds won't germinate. If they come too late, seeds will rot. Appropriate behavior and appeasement of gods and ancestors are necessary—especially Chahk and the Maize God. In addition, nothing is wasted. For example, archaeologists suggest that at the small ancient Maya community of Chan in west-central Belize, Maya added ash from hearths to nearby terrace soils to improve their fertility.[9]

Once the rains begin, the plants grow and mature quickly. In slash-and-burn fields, soil fertility lasts years before fields are left fallow for four to seven years. Fallow fields continue to provide fruit, so to speak. The secondary vegetation that grows, for example, the trumpet tree (*Cecropia peltata*), which the Maya use in house construction (and which house stinging ants), and other vegetation, provide fruits, nuts, and botanical medicines. This flora continues to feed people—and attract game. Maya also grow root and tuber crops in "fallow" fields. Spent maize stocks and other dried agricultural byproducts can be used as fuel. Nothing is wasted.

***

A map of the ancient Maya world would show a mosaic of farmsteads, fields, cities, *bajos*, and forests. As with other tropical areas, diversity also exists in terms of elevation, annual rainfall, soil types and fertility, and vegetation. Rather than being concentrated in floodplains along rivers with annually replenished soils like, for example, the rich alluvium along the Nile in Egypt, fertile soils in the Maya area are dispersed throughout the landscape in variously sized pockets, resulting in a patchwork of rich soils, seasonally inundated *bajos*, some of which the Maya also farmed, poor soils, and waterways. And farmers go where the fertile soils are. Consequently, the Maya built their farmsteads to mirror the patch-like distribution of fertile land. The larger the pocket of fertile soils, the more farmsteads there were and the bigger the cities nearby were.

Ancestral Maya farmers knew the best soils for agriculture—dark well-draining loamy soil with limestone gravel. In fact, some of the best tropical agricultural soils are found in the Maya area—mollisols. As I noted earlier, Mennonite farmers only buy land with lots of Maya house mounds because they know that ancestral Maya farmers recognized the best soils to build their farmsteads and plant their home gardens and *milpas*. However, they rarely plow around mounds. All they want to do is farm. Period. And they, at least the ones I have spoken to, have no appreciation for the ancestral and living Maya. For instance, in May 2022 at the large gleaming new Spanish Lookout Town Hall while talking with a farmer who had been negotiating between the other farmers and me about permission to excavate on their land, I was telling him about the importance of preserving Maya mounds when he asked, "The important stuff is near the bottom anyway, right?" I then proceeded to explain how the opposite was in fact the case and even pointed out the window at a random house and made the point that one could bulldoze it quickly and leave little trace of its existence. And the reason it was taking farmers so long to plow down ancient Maya houses/farmsteads was because generations of families had lived in the same places for centuries and longer.

Prior to lidar mapping, an efficient way to locate and map ancient Maya farmsteads, which archaeologists call house mounds or mounds based on how they look at present due to their having collapsed in on themselves over time, was to use soil maps. These maps can be used to characterize the agricultural capability of soils for hand-cultivation technology (i.e., digging sticks and stone tools) incorporating effective root zone, susceptibility to erosion, workability, drainage, and inherent fertility.[10] In areas with fertile soils, we found lots of mounds. In areas without fertile soils, we found few or none.

Not surprisingly, we find mounds and cities on Class I soils (fertile well-drained alluvium, e.g., Saturday Creek along the Belize River) and Class II soils (fertile well-drained upland soils, or mollisols, e.g., the medium-sized city of Yalbac near Yalbac Creek). As alluvium, however, is somewhat rare in the Maya world since most rivers are entrenched, the Belize River is a rarity with its ample alluvium. Thus, most ancestral Maya farmed Class II soils, where archaeologists find the highest settlement density. No surprise here. My project area contains less than 20% Class II soils, while Tikal's environs have over 50% (but no Class I soils since there are no nearby rivers). The Maya at Caracol compensated for their relatively poor soils by transforming the landscape with terraces that prevented soil erosion and captured water.

While some Class III–V lands might not be suitable for agriculture, they do contain plants and animals. The botanical survey revealed broom tree (*Cryosophila stauracanatha*) and cohune palm (*Attalea cohune*) in areas with poorer agricultural soils, as well as *xate* (*Chamamadorea* sp.), the palm leaves of which are used for decoration during ceremonies. Interestingly, when the large leaves of the cohune palm fall to the ground and decay, they generate black soil, a rich organic soil. I expect that Maya farmers planted cohune palms to create fertile black soils for farming.

Soils, like everything else, are living entities. The soil maps we use to locate Maya sites are just ones created by non-Maya for various reasons—to learn, use, or exploit. The Maya had their own cultural map of soils and their locations. Archaeologists just plotted what the Maya already knew for millennia, using their own soil classification systems that focused on texture, color, consistency, and stoniness, as well as potential uses. And even though the Classic Maya inscriptions largely focus on royal dynasties, epigraphers have identified several relevant glyphs that indicate the essential importance of *kab'* (earth): *lum* (soil), *luk'* (mud), and *tz'iik* (clay).[11]

The Maya perform planting and harvesting ceremonies to ensure a plentiful harvest and to thank gods and ancestors for rain, soil, healthy plants, and sunlight. Among traditional Yucatec Maya in the village of Chan Kom in the northern Yucatán, for example, the agricultural calendar makes up much of the Christian church's ceremonial schedule.[12] The life cycle of maize and other staple plants determines the cycle of daily life, rituals, and life.

Everything is connected—and acknowledged.

Plants, Animals, and "Pests"

At present, traditional Belizean fare is an amalgamation of Belize's immigration history: Maya, Spanish, British, Caribbean, and African. For example, at present, rice (non-Maya domesticate, or NMD) and beans with stew meat, usually chicken (NMD), coleslaw (NMD), and fried plantains (NMD) are the main dishes of Belize. Rice and beans are made with coconut oil (NMD) and mixed together. You need to ask for beans and rice to have them served separately. The meat, whether it be chicken (NMD), beef (NMD), lamb (NMD), venison, or gibnut, falls off the bone because Belizeans stew it in annatto or achiote (*Bixa orellana*), a crushed red seed with an amazing savory flavor. You can find some of the best BBQ in Belize—especially from street vendors, who use 50-gallon steel drums to cook and smoke their meats. Their sauces are family secrets. Then there is green maize, which one can only get in August, and *tamalitos*, tamales made with sweet maize instead of meat. *Chicha*, fermented maize, is strong stuff and should only be imbibed sparingly.

The ancestral Maya kept stingless bees (*Melipona beecheii*) for their honey, wax, and to pollinate crops, and perhaps domesticated turkeys (*Meleagris gallopavo gallopavo*) from Central Mexico, which are different from the well-known oscillated turkeys found throughout the Maya area today (*Meleagris ocellata*). Honey has different medicinal qualities depending on from which blossoms bees collect pollen. They kept tame Muscovy ducks and raised domesticated dogs (*Canis lupus familiaris*) that likely came with the First Americans for hunting and for eating.[13] They really didn't domesticate many animals. Why domesticate animals when the gods created so many "useful" ones? Today, hunters sometimes bring home the offspring of game they killed to keep as pets, especially coatimundi, spider monkeys, deer, and peccary. They don't kill or eat them, even when they reach adulthood. This practice is what we would label today as conservation because it prevents the overhunting of animals and maintains their diversity.

Many well-known domesticated and nondomesticated foodstuffs from the Americas were cultivated by the Maya—tomatoes, tobacco, avocados, pineapple, maize, beans, squash, cacao, manioc, chili peppers, and more. Where would Italian cuisine be without the succulent tomato? And we know how tobacco has impacted world health. Further, there are many varieties of these foods—chili peppers, for example, come in all colors and sizes—and degrees of heat. The smaller the pepper, the hotter it is (e.g., *maxik* or bird chili pepper, *Capsicum annuum*).

While cacao (*Theobroma cacao*) was domesticated in South America, it was the Maya who took it to new heights as an elite ritual beverage made of water, crushed cacao beans, and chili. Cacao beans also were used as a currency by some Maya. Cacao can only be grown in certain soils because it requires saturated clayey, rich soils—like alluvium. Once planted, it doesn't require lots of maintenance. There even may have been cacao estates in the past, which could account for why the inhabitants of Saturday Creek and other Maya living along the Belize River alluvium were relatively well-off (as well as not having been beholden to kings and their tribute demands) and why we find open or unsettled areas among such cities.

The plethora of domesticated and nondomesticated plants means not only that the Maya have varied fare, but also that the plants have varied qualities—some are less susceptible to disease, drought, insects, or other banes of agricultural crops. So even if some crops fail to bear fruit, so to speak, due to pests, too much water, not enough water, or other reasons, others do, providing the Maya with enough foodstuff to survive. This fertile tropical world also sustains pests. Corn smut (*Ustilago maydis*), for instance, is a fungus that grows on maize. However, it is eaten as a delicacy (*huitlacoche* or Mexican truffle) in Mexico, a culinary tradition that began with the Aztecs. Perhaps the Maya ate it as well; it is quite nutritious.

The Three Sisters play a role in controlling pests because diverse plants result in diverse insects that eat destructive critters.[14] There are also the various species of the spider mite (Acari mite family *Tetranychidae*), the corn earworm (*Helicoverpa zea*, an introduced pest), the fall armyworm (e.g., *Spodoptera frugiperda*), the corn rootworm beetle (*Diabrotica v. zeae*), and lots of species of grasshoppers, all of which cause crop damage according to American farmer Nathan Jaeger, whose family co-owns Banana Bank and who runs the agricultural side of things. He relies on Western technology—heavy equipment and chemical fertilizers and pesticides. He informed me that agricultural pests are worse in the dry season since wet conditions result in fungi that kill off pests. Dry conditions don't allow for this and promote longer and more fecund breeding of pests.

Other "pests" include coatimundi, birds, squirrels, and deer. As mentioned in Chapter 5, Maya sometimes hang a dead coatimundi to keep their living brethren away since they don't like the scent of their own dead. Scarecrows aren't efficient when you are planting large fields; they work in small ones, though. That said, for the Yucatec Maya in Quintana Roo, Mexico, "Bird and insect damage is tolerated" because "They have to eat, too."[15]

Distinct from monocropping practices, Maya farmers plant noncontiguous plots with domesticated and nondomesticated plants to prevent the spread of pests, a practice found in tropical regions all over the world. And it works. Since planted *milpas* are surrounded by nonplanted areas, if pests take over, at least they cannot spread to other *milpas*.

Relying on diverse food sources from forests, gardens, and *milpas* provided the means for the Maya to live sustainably for thousands of years. But all of this would be moot without enough rain—and at the right time.

Make it Rain

Song for the Rain:
I'll kneel for as long as it takes.
I'll stretch myself out before you on the ground,
Father Thunder, Mother Thunder.
My cornfields are suffering. The corn is drying up, Kajval [Earth God].
Hunger gnaws us with its dry mouth.

Make the Earth green.
Let the milpa flower.
Put something into our mouths,
even if it's just a bean, just a pea.
At least a lima bean, a pumpkin, or a turnip leaf.

Let the water fall from your eyes,
Father of the White Mountain, Mother of the White Cave,
Lord of the Snake,
He Who Has His Belly Full of Serpents,
He Who Eats Jaguars.

Make all the clouds from the lands come together.
Make the rain work on the whole Earth.
We want water from your three sacred wells, Kajval.
We don't want wind.
We don't want lightning.
Nor roaring thunder, nor hail.

Just water, Kajval.
to wet the dust,
to end this drought that bites us.

MARÍA XILA[16]

Everything is rainfall dependent. The most labor-intensive time for planting, weeding, and harvesting occurs during the rainy season. Maya farmer Antonio of Banana Bank told me that farmers have up to 15 days from planting when problems might arise if it does not rain when it is supposed to (e.g., seeds rot or ants eat the maize heart or the soft part). It is fine if it rains a day after planting. But if it rains too much, seeds can wash out or rot, like they did in 2008 when a tropical storm washed out all the seeds in the hundreds of hectares Nathan's crew had planted. The mud was too deep for heavy machinery to replant seeds, so his crew ended up using digging sticks to manually replant the seeds—the method the Maya have used for thousands of years.

Nearly everyone has their own way of predicting when it will rain. In mid-June in 1997 when the rainy season was supposed to start, Mr. Scott said that if it was breezy that day, which it was, that the rains would hold off. They did. We weren't so lucky in 2004 when heavy rains hit early in June. Maya crew told me it would rain on and off all day because of the new moon. In addition, frogs and toads are great indicators of rain since they mate immediately prior to the rainy season—they attract mates by croaking quite loudly all night, for several nights. Rain ceremonies often include young boys mimicking the croaking of frogs.

When rains do not come at the right time, or if there is too much rain, or not enough, the Maya pray to the gods and perform ceremonies. In return for providing the right amount of rain at the right time, the gods require appeasement and gifts—the first fruits of harvest, sacrifice, and sometimes much more. The Maya still perform the Cha'a Cháak, or "bring rain" or "request for rain" ceremony, where every man participates. The water used in the ceremony must come from a sacred *cenote* where no women are allowed to go. As recorded in the 1930s,

> The h-men [traditional healer] tells the others that it is to this cenote that the chaacs come to fill their calabashes when they are about to water the young maize plants. The cenote can be reached only by crawling through a dark and slippery tunnel about 30 meters in length. The difficulty of the entrance, and the snake-like movement of the torch-lit procession enhances the awesomeness of the ritual act.[17]

Cristina Vidal-Lorenzo and Patricia Horcajada Campos recently compiled the major features of the Cha'a Cháak ceremony based on ethnographic resources and their own field work. I summarize the main points here.[18] Ceremonies take place on an altar table, the legs of which are in the Earth, connecting the altar to the Underworld. Over the altar Maya place palm arches at the cardinal points, representing *ch'e'n*, where Chahk resides. Offerings are placed in gourds in the center of each of the arches on top of a cross. Connecting the altar to the sky and the Upperworld are vines coming from the arches. Participants offer food and drinks to the gods and ancestors (maize, chicken or turkey, *balché*) and pray for bountiful rain and crops. The ceremony ends with everyone partaking of the food and drink.[19]

Make it rain.

Farming and Migration

What does migration have to do with Maya farming? It has to do with not overusing the land, water, soils, forests, and other entities. Families grow, generation after generation. They need more land to thrive. Some farmers intensified their agricultural practices by adding terraces, dams, ditches and other agricultural features. Eventually, even this was not enough in some areas, so young families left for greener pastures, so to speak, sometimes not too far from home and other times quite far. They had to take care of their families and not ruin their reciprocal relations with nonhumans (soils, water, and forest and their denizens). This is how the interior southern lowlands became settled, as well as every other area.

This type of migration occurred in every agricultural society throughout the world. This is where strontium isotopic analysis of human teeth and bones becomes key—it traces where people were born and where they grew up, which helps archaeologists reveal migration patterns over time. Analysis of bone determines where you spent the last 10 years of your life, and teeth, the first 5 or 10.

Mobility has always been a part of Maya life for many, past and present. Planting fields that are further and further away ultimately results in families moving and establishing daughter communities, interacting with different cities. The reasons the Maya migrated should sound familiar because the same reasons exist today. Unstable climate (Terminal Classic and today's climate refugees) and economic opportunities (e.g., Postclassic maritime trade) promoted and continues to propel migration. Migration also became a strategy used to escape the labor and religious demands during the Spanish colonial period. Consequently, the Maya in any given area are not necessarily living in the same place as did their direct ancestors. Yet we find houses that were occupied for nearly a millennium (e.g., at Saturday Creek). There would not have been enough space and land for generations of offspring, resulting in younger generations departing and establishing farmsteads and communities elsewhere.

The message: the Maya did what they had to do for the survival of their families and their environment. They adapted. They made hard decisions, like leaving friends and families, to ensure everyone's continued survival—humans and nonhumans alike. Migration today is not so easy for lots of reasons—the existence of borders being one of them.

Territorial borders and tensions in the Maya region did not exist in the past to the extent we see presently. But with modern political infrastructures came territories drawn on a map and implemented on the ground, resulting in very real border conflicts. Guatemalan–Belizean tensions have existed for decades if not longer. For example, in the 1980s and 1990s border officials on both sides fumigated the undersides of each other's vehicles when crossing the border. When you enter the no-man's land between the two countries, there is a huge billboard with a map of Guatemala—that extends all the way to the coast. Belize is labeled as a district—like the Petén. In other words, the Guatemalan Government still does not recognize Belize as a country.

Guatemala finally recognized the former British colony of Belize in 1991 and established diplomatic relations. A few years later, the several thousand British troops that were stationed in Belize to protect its border with Guatemala returned to the U.K. The British armed forces had a major impact in Belize—brothels, an economic boom, girlfriends, wives, and black-market goods. Things got better between the two countries—but Guatemala is at it again, claiming over half of Belize as their own, arguing that they have a right to Spain's fifteenth and eighteenth century claims on Belize. It will have to be resolved in the International Court of Justice at the Hague.[20]

"Migration" is not a political term or concept but is something that we have been doing for millions of years, from our earliest ancestors in Africa to our first ancestors out of Africa 1.8 million years ago (*Homo erectus*). There are many reasons people move to a new place. Today, many people do so because of political instability, economic opportunities, and more recently, due to climate change. In Champaign County, Illinois, where I live, Q'anjob'al Maya started arriving in the 1980s to escape the civil strife in Guatemala.

***

Maya interaction with flora through agriculture created an environment that was abundant and allowed for species to flourish and endure, enough to feed families for millennia without massive deforestation. The lesson is clear.

Chapter 7

House and Cosmos

The ancestral Maya. People envision temples peeking out from the jungle treetops shrouded in mist. While all Maya walked amidst the mists, only a select few walked up the steep stairs to the temple summits or entered the royal palaces. This chapter, though, is about an underappreciated, humbler aspect of Maya society—house and home: the building blocks of society. More specifically, the majority—commoners who farmed and took care of their families. In fact, the "house's soul is even more powerful than the person's, since the house consists of so many elements derived from the Earth Lord (woods, mud, thatch, and so on)."[1]

When Maya first moved into a new area, they and their descendants invested in place for the long term, not just materially, but spiritually to emphasize their connection to place, other entities or beings, and the cosmos. That is, they integrated and assimilated, a reciprocal relationship reiterated and sustained via renewal. Renewal is a type of recycling, if you will, like the souls discussed in Chapter 4. Take Saturday Creek, where Maya families lived from c. 600 BCE to the early 1500s CE. Its long history would not have been possible if its residents did not live sustainably through long-term relationships with soils, water, forest flora and fauna, clay, limestone, and other nonhuman entities.

What comprises a Maya home? Typically, a thatch house with nearby smaller structures, walled or not, for storage and cooking, surrounded by a garden with an array of plants, trees, and flowers. Women cook (e.g., corn tortillas and beans) in a small *palapa* or open-sided thatched structure to keep the home free of smoky cooking fires. In the past, wealthier or elite families built their residences and shrines with mortared and plastered walls of cut limestone. Household items include ceramic storage, cooking and serving vessels, as well as *incensarios* to burn *copal* incense during ceremonies, a granite *mano* and *metate* to grind maize, chert hoes, knives, axes and scrapers, and perishable items that are invisible in the archaeological record—gourds used as pitchers, cloth made from either fig tree bark (*Ficus* spp.) or cotton (*Gossypium hirsutum*), baskets and mats, and other organic materials. Organic items just don't last in the humid tropics, but decompose and nourish soils, insects, and other entities.

Thatch roofs were and still are made of palm leaves like those of the bay leaf palm (*Sabal mauritiiformis*), which when overlapped correctly are impermeable to rain. The walls are made of narrow wood posts to allow airflow or wattle-and-daub—walls of interwoven sticks and twigs covered with clay or mud. Thatch structures are perfect for tropical environments because they are dark and provide a cool respite from the intense sun and humidity. Today's zinc roofs often leak. Nonetheless, they are slowly winning out.

After about 20 years or so, thatch roofs need replacing. By this time, the roof has become home to snakes, possums, scorpions, tarantulas, and a host of other creatures you just don't want to know about.

But the Maya house is so much more than its material composition. It embodies family history. Families recorded and updated their history through burials and ritual deposits upon which they walked, slept, conversed, and ate. Your average Maya had no need for the written word, as they experienced their ancestral history daily through stories passed down through the generations and when burying a departed member in the same spot as earlier ancestors had been for decades if not centuries.

The Maya home is the cosmos writ small. Houses embody the three levels of the universe. As mentioned in Chapter 4, the roof is the head, the space under the roof is the sky and the Upperworld, the walls are the mountains and/or the stomach, the floor the Earth's surface and/or the foot, and underneath the floor is the Underworld. The family is a branch of a tree, signifying lineage and connections to the sky, Earth, and Underworld. Everything is connected. The house is part of the landscape, enmeshed with other nonhuman entities expressed through transformation and renewal—from wood and limestone to house, clay and temper to ceramics, chert to stone implements, ceramic vessels to ritual deposits, life to death, and death to life. This cycle propels a sustainable existence that we can see in the archaeological record.

Transformation and Renewal

So the New House Won't Eat Us:
My house is built,
my hearth is made,
Three Sacred Mothers,
Holy Earth, Kaxil [Earth Goddess]
We are going to live here.
We are going to sit here.
We are going to shit here.
We are going to pee here
beneath your flowering face,
each day, beneath your flowering eyes.

I am touching the ground with my forehead
as I say this, Kajval [Earth God]

We are going to sleep here.
We are going to rest here.
We are going to sin here
and make love.

I hope your heart won't scold us,
that your blood doesn't nag us,
that nightmares don't disturb our sleep here,
nor witchcraft.

The neighbors claim we are rich, Kajval,
but we have no treasures.
Close the mouths of the envious
so they can't gossip about us.

Let the saints in the coffers,
the voices that speak from inside old chests,
stand up for those who live in this house.

Protect us from being eaten by a vine or a stick.
Save us from being devoured by the new thatch
or the shiny nails.

Here is your food, Kaxil.
Dinner is served.
Eat this meat
instead of the bodies of your children.

We offer you gifts, Kajval,
so the new house
won't eat the people in it.

XUNKA' UTZ'UTZ' NI'[2]

As this animating prayer shows, houses are living entities, ensouled when families perform animation rites. Before they rebuild a house, the Maya perform de-animation rites. First, a funeral rite acknowledges the death of a loved one, as well as a new beginning. The family then destroy and burn their house and household items as part of the de-animation ritual to release their *ch'ulel* (soul) before the remnants of the burned and broken things are animated as the foundation for the new house. The new house, or more accurately, the renewed house, rises from the ashes of the former one, which becomes ensouled via animation rites. Renewal and construction are thus one and the same. Houses thus entomb family histories, allowing for children as *k'exol* or replacements of their ancestors to walk among their deceased elders, tethering the younger generation to the old. They, in turn, add to the family history upon their death.

The Maya did not separate the landscape into cultural and natural categories. Rather, they added to the landscape. Once permission was granted by the Earth Gods to use their materials, they were transformed from one animated form into another through rituals and offerings. A house then became a home, ensouled, and a microcosm of the world, renewing a family's balance with the cosmos. The history of building, destruction, and rebuilding is thus a history of transformation and renewal. We see the remains of this cycle in the archaeological record via depositional histories, that is, layers upon layers of construction events, burials, and de-animation and animation deposits. In fact, the material manifestations of rituals are a large part of the archaeological record. The Maya kept ancestors close, burying certain deceased family members underneath house floors along with funerary objects. They then de-animated everything by smashing and burning them. To animate and ensoul the new house, Maya dug holes or portals into the floor, placing offerings and burying them.

The Maya often inverted vessels over the deceased in their final resting place underneath the de-animated floor surface or placed lip-to-lip vessels. As mentioned in Chapter 4, lip-to-lip vessels also signify the cosmos—the empty space under the top vessel is the Upperworld. The Earth is reflected in jade, hematite, red ochre, cacao, foodstuffs, or stones contained in the bottom vessel. Sponges, stingray spines, shell, or coral signify the sea. The Underworld is indicated by nine black flakes signifying its nine aspects. Inverted vessels signify death and renewal or rebirth.

The Maya also animated other materials provided by the gods, transforming them into items for daily use: ceramics of all forms, backstrap looms for weaving textiles, woven baskets and mats, stone agricultural implements such as chert hoes and axes, granite *manos* and *metates*, and stone tools for cutting, hunting, and fishing.

Maya rituals have changed little over the millennia, as archeological, ethnohistorical, and ethnographic cases show, because they were vital in world maintenance through renewal. Today, traditional Maya *curanderos* (healers) continue to perform "new house" or animation ceremonies throughout the Christian and non-Christian Maya world. The Tzotzil of Zinacantán, Chiapas, Mexico, for instance, perform two stages of this ritual, one during and one after house construction. During, they bury the heads of chickens in the center of the floor along with other offerings. After, the *curandero* performs rites to compensate the Earth God for the materials he provided and to "summon the ancestral gods to provide the house with an innate soul."[3] More offerings are then buried. The accompanying prayers can include references to the Christian God, the Virgin Mary, or Jesus Christ, as witnessed among the Tzotzil in Chiapas, Mexico.[4]

The sheer number of offerings in the archaeological record is staggering, especially ceramics. For instance, project ceramicists Jennifer Ehret and Jim Conlon analyzed nearly 6,000 sherds from Saturday Creek in 2001—and this number does not include those they could not identify (e.g., plain body sherds). Despite the millions of ceramics recovered from southern lowland Maya sites, we have yet to identify an obvious ceramic manufacturing site; that is, their production left a minimal footprint. Is this an example of sustainable transformation? I think it is. In fact, ceramic production provides an excellent example of how ancestral Maya families left a nominal footprint even though they manufactured literally millions of ceramic vessels, figurines, net weights for fishing nets, spindle whorls for cotton spinning, and a host of other ceramic items. Let me explain.

Interlude: Millions of Ceramics

In this section, I ask you to think like an archaeologist: What kind of evidence would we find in the archaeological record of ancient Maya ceramic production? Here are some clues.

In 1989 I visited Amatenango in Chiapas, Mexico, a Tzeltal Maya village famous for its ceramic *animalitos* (miniature animals) and vessels. We stopped at the first pottery-producing household we came upon. The potters allowed us to take photos if we bought something. So we did. The lady of the house was applying a white slip, a mixture of clay and water, on a sun-dried, leather-hard *olla* (jar). A younger female family member was burnishing a slipped olla with a river-polished pebble. The pigments used to make brown, mustard yellow, and burnished red paints are ground minerals of some sort mixed with water in halved chert nodes. Another young woman was painting decorations on unfired, leather-hard pig and armadillo banks for the tourist market using a fibrous brush made of cane frayed at one end.

Another pottery-producing household consisted of five women potters, each having a specific task. Of the five, two sisters were the main potters. They were kind enough to share their knowledge. One of them was burnishing a slipped *olla*, one of the many she and the other women in the family (men work in the fields) were contracted to make for a client in Oaxaca over 650 km (over 400 miles) away. One of her sisters was forming the rim and adding the neck and handles to an *olla* (Figure 7.1). She could easily turn the *olla* because it was resting on a thin layer of ground temper (crushed limestone in this case) with the consistency of sand—a type of "potter's wheel." She used the distance from her extended forefinger to her thumb to measure where to place the handles. They then smoothed the *olla* surface with a metal rasp (scraper), after which they added slip and then burnished with a river-polished pebble. The point of presenting these details is to highlight the traditional techniques and tools used to manufacture ceramic items—and the

FIGURE 7.1 *Traditional ceramic production practices in Amatenango, Chiapas, Mexico, 1989*

In the upper left, the ceramicist is measuring the distance between the handles. Note that the jar rests on a thin layer of sand-sized ground temper to make it easier to turn. In the upper right, a household member polishes a sun-dried vessel using a smooth pebble; in the lower left, another uses a flayed stick as a paint brush. On the lower right are young village girls selling their *animalitos* (miniature ceramic animals).

Source: Photos by the author, taken with permission.

implications for finding evidence for ceramic production in the archaeological record.

The clay and slip came from a source about a two-hour walk away. I don't know where the temper came from. Potters from several households fire every two weeks on the street in an open-air fire using old rubber tires for fuel.

Potters ask permission of the Earth Gods before making use of their clay.[5] Traditional Maya potters of Jocopilas, Guatemala, for example, walk to sacred places in the mountains and perform rites to ensure success in pottery manufacture "to maintain a balance with the elements of the universe.... through the ritual, the potter relates to the universe in submission and feels herself a part of all life."[6]

As with nearly everything else, pottery production, distribution, and ceramics have animated qualities. Lacandon Maya potters of Chiapas, Mexico, for instance, perform animation and de-animation rites for vessels. To animate them, they say prayers. For god pots (*incensarios*), they also offer rubber figurines in the form of humans to symbolize flesh and blood. God pots, when replaced, must be de-animated and buried.[7] A few days later the old god pots are taken to a cave and left there (see Chapter 4).

Why is identifying ancient Maya ceramic production in the archaeological record so challenging? Let's take stock.

First, the Maya did not use potter's wheels in the southern lowlands. They used hand coiling and slab building. Nor did they use kilns, but instead used open-air fires. With thatch houses and plastered buildings susceptible to destruction by fire, ancestral Maya potters would have had to fire ceramics at some distance from buildings. So, unless we just happen to chance upon huge fire pits in nonsettled areas, we would not find them near ancient Maya constructions. In the eastern Puuc area of the northern lowlands, however, archaeologists have found evidence for ancient fuel-efficient kilns compared to colonial and modern ones, indicating that precolonial Maya collaborated with their nonhuman neighbors in such a way as to avoid environmental degradation.[8]

Second, local clays and temper sources are readily available and widespread, so we don't find a community of potters who live in a specific area due to prime clay and temper sources per se. The predominant clay type is calcareous and fuses at relatively low temperatures. This montmorillonite clay is also quite sticky, a characteristic that can result in high shrinkage rates during drying and firing, and thus, may result in cracking. Adding temper offsets this problem. Ancestral Maya potters used three major temper types: limestone, volcanic ash, and grog (crushed or ground fired pottery).

Third, many wasters (melted ceramics or what's left of vessels that exploded during firing) are hard to find due to the lower temperatures of open-air fires compared to kilns and the tempers potters added to clay. We just don't find many wasters in the archaeological record. To figure out why this is the case, I conducted a ceramic firing experiment using modern and ancient Maya sherds and emulated open-air fires in kilns at the UCLA Art Ceramics Laboratory of the Department of Art.[9] Open-air fire temperatures fluctuate (625–900+ degrees Celsius) and require skill to control. If there is too much of a draft, vessels break. If temperatures are too high, vessels can crack, shatter, or melt.

When production failures happened when firing limestone-tempered vessels, as invariably they did, they would basically be invisible in the archaeological record because broken vessels look exactly like regular sherds. Further, the experiment demonstrated that misfired limestone-tempered vessels crumbled into sand-sized bits once cooled, which ancestral potters probably recycled as grog temper. Volcanic ash–tempered vessels can withstand high temperatures. Thus, any misfires would basically be indistinguishable from sherds resulting from vessels broken during use.

Fourth, ancestral potters relied on perishable or indistinct tools used to shape and prepare vessels. Chert flakes and gourd pieces were likely used as scrapers to smooth the coils or slabs into a seamless vessel. Plant stalks were fashioned into paint brushes. They used river-polished pebbles to burnish vessels. They used a thin layer of sand or crushed temper on the floor surface to assist potters to turn vessels. They used rocks with natural depressions, ceramic vessels, and gourds as containers for the organic or mineral paints used to create beautiful designs. We will not find gourds or anything organic in the archaeological record. And even if we find sand, flakes, halved chert nodules, and river-worn stones, there is no way to prove that they were used in ceramic production unless we find substantiating evidence.

Fifth, traditional potters today manufacture ceramic jars, plates, dishes, bowls, figurines, and other forms at the household level—that is, small-scale production—as did past Maya potters for local or regional exchange. Thus, the potentially thousands of production loci are dotted all over the Maya area.

Ancient Maya ceramic production thus left little evidence in the archaeological record. Simultaneously, it did not leave an obvious footprint, a surprising fact given the huge numbers of vessels and sherds archaeologists have recovered—and the millions more from unexcavated sites.

Renewal of House and Cosmos

Having excavated many ancient Maya houses over the years, I have exposed layer upon layer of burials and ritual deposits interspersed with floors and walls, especially at Saturday Creek (c. 600 BCE to early 1500s CE).[10] Saturday Creek is a small city that sits along the Belize River with a high density of commoner residences and elite *plazuelas* (two to five structures around an open area or plaza) surrounding a ceremonial core consisting of a ball court and temples up to c. 10 m (31 ft) tall (Figure 7.2). Its plentiful resources, including alluvium, an aquifer, a major river and transportation route, and nearby forests, meant that any political aspirants were kept in check since they could not provide anything the local Maya could not attain themselves. So, no kings at Saturday Creek. There are other comparable cities without kings, such as Barton Ramie to the west, also along the Belize River.

Excavating ancient Maya houses is time-consuming. Why? Because they are chock-full of plaster floors with de-animation deposits on their surface and animation ones and burials underneath. Their depositional histories signify detailed and long family histories

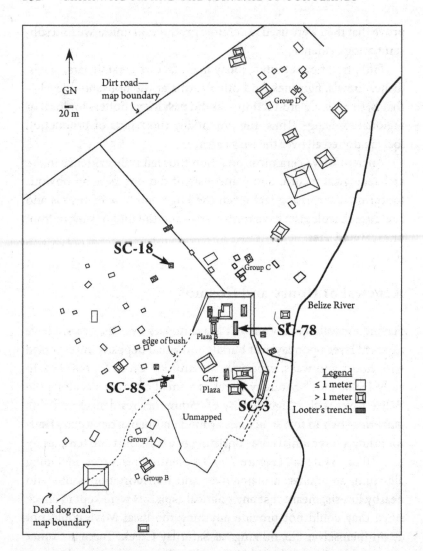

FIGURE 7.2 *Saturday Creek's ceremonial core, farmsteads, and elite* plazuelas, *including excavated sites*

Source: Courtesy of VOPA.

showing the longevity of their sustainable relations with nonhuman entities. The same goes for other ancient houses throughout the Maya world.

Offerings are not haphazardly placed. The color of objects (red versus black, for instance) and their positioning tell their own stories. The Maya sometimes positioned caches in such a way as to divide the home into quarters, highlighting the four cardinal directions and center. Often, they placed offerings in groups or layers of three to signify the three levels of their universe—including the Underworld, which they accessed when they dug into floors and left offerings and interred the dead. From commoners to royals, people offered *metates* and *manos*, signifying the never-ending essential importance of maize. There was an entirely different world on which the Maya walked, but did not see, day after day, year after year, until the renewal cycle began again.

Burials containing deceased ancestors established place. Years later, family members sometimes removed body parts for various reasons: to make way for another burial, to keep as keepsakes in sacred bundles, to rebury with other ancestors, or to include as offerings like the human caches discussed in Chapter 5. The Maya knew burials were underneath their feet, even centuries later. Family histories and ancestors were never erased from memory. Today, the "souls of the dead work for the Earth Lord and visit their families once a year on the Day of the Dead" (November 2).[11]

The Maya did not bury every deceased family member in the home. If they had, we'd find hundreds of burials in every house. For instance, imagine if the families of the two small houses we excavated at Saturday Creek interred all deceased family members over a period of nearly 800 years. If we conservatively estimate that each household housed five people at any one time, with a new generation every 20 years, this would have resulted in 188 burials over the nearly 800-year occupation period. However, we excavated only 12 burials from both residences. In an innovative study using tomb inscriptions, ceramic chronology, and radiocarbon dating, Diane and Arlen Chase, who have been excavating the large city of Caracol in Belize for decades, estimate that the Maya buried only about 10% of the dead in homes, shrines, and tombs.[12] They further posit that the major factor determining who was buried in the home, temple, or shrine was calendrical—namely, who died closest to either the 52-year calendar round or 2 *katuns* (40 years), or in the case of thatch homes, once every *katun* (20 years).

Commoners would have buried family members who had died closest to the end or beginning of every *katun* when thatch roofs needed replacing anyway. Death, de-animation, renewal—again and again. But where did they bury the remaining 90% of departed family members? I address this question in Chapter 10. But I will say here that their corporeal remains nourished and renewed the tropical forest and its denizens while their souls continued their eternal journey—being recycled generation after generation.

Archaeologists distinguish the type of ceremonial offering based on whether items, especially ceramics, are nonblack and complete (funerary); broken, incomplete, and burned (de-animation); or whole (animation). We don't often find black vessels in burials because of what they signify—death. Instead, we typically find red ones. Red signifies east and renewal, the sun, and life. The Maya kept some broken pieces from de-animated items for other purposes, while some were deposited away from the home. For example, for the Tz'utujil Maya in Kankixaja, Guatemala, at the end of the *Tzolk'in* (order of days) or 260-day ritual calendar, family members break the main jar in which they boil maize and place the larger broken pieces on a large pile of ceramic sherds (from years of depositing broken ceramics) at an ancestral shrine in the mountains.[13]

The Maya performed the same ceremonies in houses and palaces and every type of structure in between, but writ large in larger structures, as we see when comparing two small residences (SC18, SC85) and a community-built temple complex (SC3) at Saturday Creek to a palace complex (North Acropolis) at Tikal in Guatemala (c. 600 BCE to 900 CE), the capital city of a powerful kingdom (Table 7.1 and Figure 7.3).

TABLE 7.1 Saturday Creek (SC) and Tikal structure dimensions

City and type	Size	Height	Occupation history
SC18, house	10 × 8 m (33 × 26 ft)	1.2 m (4 ft)	400–1150 CE*
SC85, house	6 × 4 m (20 × 13 ft)	1.3 m (4.3 ft)	400–1150 CE*
SC3, temple complex	38 × 20 m (125 × 65 ft); pyramid structure on east side, 5 × 5 m (16 × 16 ft)	5.4 m (18 ft) total; terraced platform, 3 m (10 ft); pyramid structure, 2.4 m (8 ft)	300 BCE to 1500s CE†
SC2, ball court structure	20 × 12 m (65 × 39 ft)	3 m (10 ft)	700–1500s CE
Tikal's North Acropolis	100 × 80 m (328 × 262 ft)	40 m (131 ft)	600 BCE to 900 CE

* Ten years of plowing these mounds erased hundreds of years of history; they were likely occupied through the 1500s.
† Ceramics dating to 600 BCE were found in lower fill deposits, indicating that the Maya occupied this area earlier than 300 BCE.

Maya families renewed the two small residences, again and again with minor additions a few inches thick (Figure 7.4). We exposed layer upon layer of thin plaster floors with pebble ballasts (floor support) and low foundation walls for thatched wattle-and-daub houses that date from at least 400 through 1150 CE. The Maya likely lived at both until the early 1500s; we will never know because their houses have been plowed for 10 years or more.

The Maya at Saturday Creek built and made moderate additions (20–65 cm, or 8–26 inches, thick) to a temple complex (SC3) for over 2,100 years, resulting in a terraced platform with plastered steps with a pyramid temple on top of the east end of the platform. Sometime during the Late Classic (c. 700 CE), the Maya added SC2, and in so doing, turned SC3 from a temple platform into a temple-ball

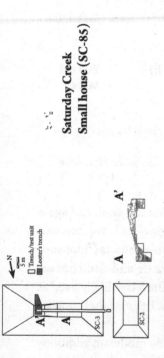
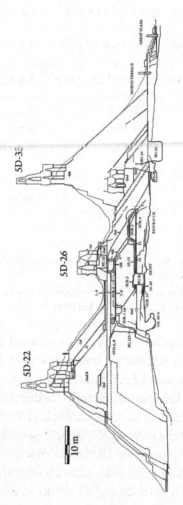

FIGURE 7.3 *This image shows Saturday Creek building profiles to the same scale as the North Acropolis profile*

The small house is barely perceptible compared to temple and palace complexes. But they all have the same stratigraphy or depositional histories reflecting renewal cycles.

Source: Saturday Creek image courtesy of VOPA. North Acropolis image courtesy of the Penn Museum, image #67-5-113.

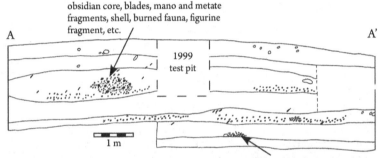

FIGURE 7.4 *Photo and profile of a small Saturday Creek residence (SC85)*

While small, it is packed with renewal cycles. Bs indicate burials. The surface artifacts in the lower-right corner are a de-animation deposit.

Source: Courtesy of VOPA.

court complex, with the area between the two structures serving as the ball court alley. Excavations exposed 14 major strata, or layers, that date from about 300 BCE to the early 1500s CE, representing major rebuilding events and many minor ones. Due to time constraints, we were unable to completely reveal all the layers, so we are not sure how deep the cultural layers go or when the Maya first began living here. Excavations at the bottom yielded ceramics dating to c. 600 BCE. Communal labor parties organized by local elites maintained and rebuilt the temple complex and performed the necessary rituals.

In contrast, by the time of its abandonment by 900 CE, the North Acropolis at Tikal had grown into a massive palace complex (Figure 7.5). It began soon after 600 BCE as a small platform about the size of the small houses we excavated at Saturday Creek—6 × 6 m (20 × 20 ft) and c. 1 m (3+ ft) tall.[14] By 250 CE, Tikal's kings were building increasingly labor-intensive additions up to 15 m (49 ft) thick that included ornate tombs, de-animation deposits, and animation caches. By the Late Classic period (c. 600–800 CE), when

FIGURE 7.5 *This image shows what remains of Tikal's North Acropolis*

During excavations in the 1960s archaeologists removed much of the overlaying architecture that later Maya kings had added. It is still impressive, nonetheless.

Source: Photo by the author.

Tikal's kings were at their most powerful, they added additional buildings to limit access.

The North Acropolis at Tikal and the temple complex at Saturday Creek started out the same but end up quite differently—40 m (131 ft) tall compared to 5.4 m (18 ft) tall. Family members built the houses. Community members built the small temple complex. Tribute labor built the palace complex. They all, however, signify the same thing—cycles of renewal through destruction and rebuilding.

Renewal at Saturday Creek

Let me take you back to a time before Columbus set foot on the shores of the Americas, when Maya still lived at Saturday Creek. As I just showed, temples and other such buildings grew incrementally over the centuries via renewal. Elites lived in bigger and bigger houses. And while everyone performed the same household rituals, elites proffered more expensive and exotic items like speleothems from caves, coral from the barrier reef, monkey finger bones, mica from the Maya Mountains to the south, and beautifully decorated vessels and larger jade items, demonstrating their success as farmers to their neighbors, gods, and ancestors. No matter how wealthy they became, however, they always maintained the vital and mutually beneficial relations with nonhuman entities that allowed them, their families, and their descendants a long history at Saturday Creek.

We excavated a small portion of a large platform (SC78: 29.4 × 9.5 m and 3.9 m tall, or 96.5 × 31 ft and 13 ft tall) and several of its structures that made up part of a *plazuela*. Elites lived here from c. 600 BCE through the early 1500s CE, constructing walls of cut-stone blocks and nice plaster floors, as well as some wattle-and-daub structures. We did not find burials in our minimal explorations here—nor at the temple complex for that matter. We did recover a shaped and polished chert flake from the south edge of the platform from a de-animation deposit dating to sometime after 1150 CE. Its blue, white, red, and black striations are associated with the cardinal directions, life, death, and things we will never know (Figure 7.6). Its

5 cm

FIGURE 7.6 *SC78, multicolored polished chert flake*

Source: Courtesy of VOPA.

polished surface shows that the Maya handled it frequently, likely passing it down through the generations.

As the Maya did elsewhere, the Maya at Saturday Creek de-animated buildings by smashing and burning vessels and leaving most of their remains on floors that they also burned. Family members de-animated vessels by breaking off their rims—as keepsakes or to deposit in a portal or other significant places. They also deposited heirloom rim sherds centuries older than the rest of the deposit. Families passed heirlooms down through the generations and waited for the right time to ceremonially cache them. Sherds or pieces of vessels are just as significant as the vessel itself and the context that it represents—family history, pilgrimages, or community and royal ceremonies (see Chapter 5).

Elites used cylindrical drinking vessels to drink a ceremonial cacao drink. In c. 800–900 CE at SC78, they performed a de-animation rite during a tumultuous time (droughts). The elite family stacked three cylindrical vessels after having had removed their rims, one inside the other. The top and bottom rimless vases were black, and the middle one was decorated. They also burned a wattle-and-daub structure in another area of the platform. Before setting the fire, they de-animated it by placing burned and smashed decorated but rimless vessels, including a burned rimless plate dating to c. 300–100 BCE, on top of which they placed a human ulna (arm bone), an incised marine shell pendant, and an inverted black miniature jar. We will likely never know the complete translation of this deposit and the history it embodies.

The temple (SC3) has its own story to tell. Maya elites sponsored ceremonies on the temple summit at a circular altar (c. 1 m, or 3+ ft, in diameter), on its associated terraced platform, and in the ball court alley for all to witness, and perhaps participate. At the pyramid temple itself, the Maya burned corozo palm nuts to de-animate an earlier construction event before rebuilding on top of it. And due to its importance to the community and sponsorship by local elites, its animation caches differ from commoner residences—monkey finger bones, ceramic balls and decorated polychrome ceramics, quartz, mica, and notched obsidian blade fragments. And in a consistent series of events at the foot of the temple along the central axis, the Maya burned items and placed inverted, broken, and typically rimless vessels. For instance, on one of the upper surfaces the Maya had placed an inverted Belize Red dish with a partial intact rim that dates to 800–900 CE. In another de-animation event dating to the same time, the Maya placed an inverted Platon Punctate rimless plate, a ceramic type not found anywhere else at Saturday Creek. In a final example from the platform, the Maya deposited, likely as part of an animation ritual, chert cores, obsidian blades, and marine shell dating to c. 900–1150 CE. The Maya performed ceremonies and feasting events in the ball court alley as well, based on the presence of a mosaic jade piece, obsidian blades, burned organic materials, burned animal bones (e.g., deer) and turtle carapaces, bowl sherds (serving vessels), and ceramic sherd clusters.

Between c. 700 and 800 CE, to de-animate one of the commoner residences (SC85), its family placed an arrangement of 10 clusters of burned and broken, rimless reddish and black vessels, some dating to c. 300 BCE (see Figure 7.4). They placed four of the clusters around six orange/reddish clusters c. 30 cm (12 inches) apart set at the cardinal directions. In its center, they added white stones, a green serpentine oval disk, and a *metate* fragment. Green represents the center of the universe, white represents rain, red represents renewal and life, and black represents death. The *metate* is associated with the sun and maize.

Burying family members also is a complex ceremonial process. For a funerary rite for an adult male family member (Burial 5) sometime between c. 500 and 600 CE at one small house (SC18), the Maya dug a pit into the floor in which they placed broken pottery and chert flakes and then burned everything. They then put a deer antler and a pink-red quartz stone on top, added dirt, and then placed the body, on top of which they added more dirt, followed by the smashing and burning of more vessels. There's more. They placed *mano* and *metate* fragments and more ceramics near the body and added more dirt. Finally, they placed whole and complete vessels and other items south of the skull and then burned the entire deposit. All these ceremonial stages would have been accompanied by chants, prayers, music, feasting, and burning incense. But this still was not the end of it. The Maya then razed and burned the wattle-and-daub house and performed a de-animation rite placing clusters of inverted smashed and burned vessels. Then, of course, the family rebuilt and animated their new home. Everything was renewed, and balance was maintained.

Updating family history often meant interring deceased family members in the same spot for centuries through several rebuildings. This is probably the only time the Maya saw "stratigraphy" in the sense archaeologists do. For example, at c. 700 CE at one of the small houses (SC18), the Maya interred an adult female family member (Burial 7) by laying her on her back and inverting a red bowl over her knees along with freshwater shell disc beads. At c. 750 CE, her descendants removed her upper torso and skull and interred an adult male family member sitting cross-legged (Burial 11), with a large incurving dish inverted over his head. They then placed her long bones in front of the seated male, reiterating family ties and stories only they knew. We did not find the rest of her remains, which her descendants removed—the family either kept them as keepsakes, reburied them with other ancestors, or offered them in-house or elsewhere, such as a pilgrimage destination.

After this funerary rite, the Maya performed a complex de-animation ceremony. They positioned broken ceramics at cardinal directions on top of some burned textiles or mats on which all that remains is ash and a burnt rectangular stain. The Maya first placed heirloom ceramic sherds dating to c. 400–600 CE over the burials. They then placed more sherds, followed by another layer of decorated and inverted ones. Now it was time to rebuild and start the cycle again.

In 20 or so years, the renewal cycle would start all over.

***

The different scales of ritual deposits at Saturday Creek, the North Acropolis at Tikal, and elsewhere in the Maya area show how everyone in society acknowledged reciprocal relations with each other, ancestors, gods, and fellow entities because they all lived in and took care of the same world, resulting in a long-term, sustainable existence.

The Maya transformed other entities to survive, and in the process, changed relations without detrimental consequences for millennia. It began in the home for everyone from commoner to king, and the home itself became part of the cosmos.

Chapter 8

Water Lily Kings

When farmers move into new areas empty of people, founders or first comers soon distinguish themselves by building larger houses and acquiring more goods, especially imported ones. While everyone performs the same traditional ceremonies, first comers and their descendants perform more ornate ceremonies and deposit more expensive offerings. Nonpolitical elites as patrons sponsor ceremonies and feasts for the entire community. Eventually, leaders and later kings showed their might through successfully supplicating gods and ancestors via traditional ceremonies writ large. In doing so they not only demonstrate their power over others but integrate more people-slash-subjects and promote solidarity since everyone performs the "same" traditional rites.

This scenario has occurred again and again across time and space. But how did wealthy elites become political leaders; that is, when did they acquire the rights to the goods, services, and labor of others? Part of the answer comes down to supply and demand. When commoners or farmers can supply their own food and have access to all essential resources they require (trade goods, water, and forests for game, construction materials, food, and plants), like we see at Saturday Creek, they do not need top-down help—especially since there's always a cost involved. Elites, by definition, own more or at least have access to more resources that they can use to help their brethren in need. But as we all know, politicians, emerging or established, never do anything for free.

Political systems require more resources (i.e., surplus) as they expand and incorporate more people, which often translates as intensifying agricultural or comparable practices. Leaders provide capital generated from their ownership or control of resources and tribute to build and maintain infrastructure, including public facilities, water systems, roads, protection, armies—and, of course, food in times of need.

Political systems come and go. The more emperors, pharaohs or kings depended on surplus provided by others to support their political system, the more vulnerable they were to changing conditions. When problems arose such as failed crops, famine, or inclement weather, subjects blamed rulers. When a political leader fell from grace (e.g., Chinese emperors losing the mandate of heaven), elite and commoner subjects continued their daily existence—typically under a different political umbrella.

Ancient tropical political histories followed this trajectory, including Classic Maya kings.

Royal Flower Power

Ancient tropical rulership in Sri Lanka, Cambodia, and the southern Maya lowlands had major features in common—seasonality, annual rainfall averaging c. 1,500 mm, humidity, and rainfall-dependency. They also depended on reservoir systems to maintain their subjects and political power. The largest Classic Maya cities ranged from c. 100 to 200 km^2 (77+ square miles), compared to the Sinhalese Buddhist capital of Anuradhapura in Sri Lanka (fourth century BCE to eleventh century CE) at c. 500 km^2 (193 square miles) and the Khmer capital of Angkor in Cambodia (ninth to sixteenth centuries CE) at c. 1,000 km² (386 square miles).[1] Angkor and its environs supported c. 750,000 people at its height of power in the twelfth and thirteenth centuries CE. There were as many as 250,000 people living in Anuradhapura and its environs and 60,000 to 80,000 in Tikal. Anuradhapura and Angkor served as political capitals, in contrast to the largely independent Maya cities. Here, I briefly compare the Khmer of Angkor and the Maya of Tikal.

Between the ninth and eleventh centuries CE the Khmer constructed and expanded *baray* (artificial reservoirs) in the urban core of Angkor that were fed by extensive canal systems extending from several major rivers (e.g., Siem Reap River). The West Baray, for example, could hold over 15 million m³ (3,962,580,785 gallons) and measures c. 8 × 2 km and is 2–5 m deep (5 × 1.2+ miles, 6.6–16.4 ft deep). Canals drained water from reservoirs when necessary, which fed the extensive rice fields. Rice was the main staple for the Khmer, while maize, beans, squash, and manioc were the main staples for the Maya. In contrast to the Maya, Southeast Asian societies had access to metals, beasts of burden (e.g., cattle), and an extensive international maritime trade network.

For most of its history, the official religion of the Khmer empire was Hinduism. A few forms of Buddhism also had followers (e.g., Jayavarman IV, 1080–1107 CE, practiced Mahayana Buddhism), but its influence was minimal until after the Thai (Siamese) invasions in the fourteenth and fifteenth centuries.[2] Reservoirs were constructed next to palaces of wood and temples of stone. Their sunken walls were covered with carved scenes from Hindu creation history. According to inscriptions in Old Khmer, Ta Prohm, a late twelfth century temple in Angkor, had 12,640 priests, *apsara* (dancers), and support staff who were supported by 66,625 surrounding farmers who owed them labor and a portion of their rice crop. At Tikal, public plazas could hold over 10,500 people, who participated in royal ceremonies and feasts sponsored by kings with foodstuffs and accoutrements supplied by royal coffers and tribute from these same participants.

Water was a major element in their worldviews, a fact reflected in their respective iconographic records.[3] For example, at the headwaters of the Siem Reap River in the Kulen hills northeast of Angkor, the Khmer carved on the Kbal Spean sandstone river and creek beds scenes from the Hindu creation history. One of these carvings depicts the lotus flower (*Nelumbo nucifera*) coming out of Vishnu's navel, from where Brahma emerged, seated in the center of another lotus flower.[4] Gods thus received perpetual libations from waters emerging from the Earth. Vishnu's consort, the Goddess Lakshmi, is also surrounded by lotus symbolism in Hindu iconography. Her divine seat is a pink lotus, and she holds a lotus in her right hand. The lotus is also essential in Buddhist symbolism. For example, Buddha was supposedly born on a lotus leaf.

The presence of lotus flowers and water lilies (*Nymphaea ampla*) on the surface of the Khmer *baray* and Maya reservoirs denotes clean water and symbolizes purity and power. Both grow in water bodies 2–3 meters (*c.* 7–10 feet) deep.[5] The Sanskrit word *Padmasana* means 'lotus seat,' which refers to throne of Shiva, the supreme Hindu God, as well as royal thrones in South and Southeast Asia.[6] Water lilies symbolize Maya kingship. In Maya inscriptions, kings are referred to as *Nab Winik Makna* or Water Lily Lords, and Classic Maya nobility as *Ah Nab*, that is, Water Lily People.[7] Further,...

> several dynasts adopted titles that identified them directly with the flower, such as 'Sun-face Sacred Lord Macaw Water Lily' (Kinich Ahkal Mo' Naab), Sun-face Water Lily Flower Shield' (Kinich Janaab Pakal), 'Lady Water Lily Sprout' (Ix Kook-Naab) and 'Lady White Sprout' (Ix Sak Kook) of Palenque, or Water Lily Jaguar (Balam-Nan) and 'Smoke Water Lily Kawil' (K'ak-Naab-K'awil) at Copan.[8]

The iconography is replete with kings and water lilies and related elements on monumental architecture, *stelae*, murals, and portable items. Maya kings impersonated some kind of water deity depicted

FIGURE 8.1 *Maya cylindrical vessel (700–800 CE) from Guatemala that shows a king, Ho-k'in-bat, sitting on a throne wearing a water lily headdress*

Source: Courtesy of the Boston Museum of Fine Arts (www.mfa.org).

"as a serpent with a water-lily bound to its head."[9] Maya kings associated themselves with the power of water lilies as did ancient Egyptian pharaohs and Chinese emperors (e.g., Ramses II and other pharaohs were buried with water lily wreaths).[10] In fact, Maya kings wore water lily headdresses (Figure 8.1). Finally, these flowers may have been used as a hallucinogen during royal and other ceremonies. Botanists consider *N. ampla* a substitute for opium, producing mind-altering episodes. It contains nupharidine with aporphine-like or opiate-like alkaloids in the rhizome, including apomorphine, nuciferine, and nornuciferines.[11] The use of the lotus as a mind-altering drug is well documented. In India, an ecstasy-inducing drink prepared exclusively from the lotus plant is documented in the Vedic texts that date from between c. 1500 and 1200 BCE.[12] Buddhist monks also consumed a drink made with the crushed rhizomes of lotus plants.

As long as kings performed their duties well, their power was secure and the loyalty of subjects intact. However, in both cases rulers lost power—in the ninth century CE for Maya kings and the sixteenth century CE for Khmer rulers, largely triggered by long periods of climate instability when too much water or not enough of it interfered with food production and clean water supplies.[13] In Angkor, there were periods of both extreme droughts and heavy monsoons—that is, too little and too much rain that ultimately resulted in a gradual urban diaspora. Most Khmer moved to small towns throughout Cambodia, such as near Tonle Sap Lake, the Mekong River, and elsewhere. The political elite moved southeast to the Phnom Penh area, eventually making Phnom Penh its new capital by the sixteenth century, which it remains today. We know what happened to Maya kings in the southern lowlands (see Chapter 2).

Maya Kings: A Glimpse

The time: 500 CE at the end of the dry season. The place: Tikal, Petén, Guatemala.

The pipes played and the drums thundered and then ceased. The crowd filling the vast plaza and the dancers adorning the stairs of the lofty pyramid temple turned their eyes to its highest ledge.

Descending from the doorway of the temple crowning the *Mundo Perdido* (Lost World) pyramid stood a man dressed as a god. Resplendent in his vibrantly colored god costume was Chak Tok Ich'aak II, the king of Yax Mutal (First/Green-Blue Bundle), who reigned from 486 CE until his death in 508 CE. Any part of his body not covered by cotton and feathers was covered with tattoos and jade. Flanked by the life-sized masks of the Rain God Chahk, Chak Tok Ich'aak II raised his arms. He wore a blue-green iridescent quetzal feathered headdress and clutched a staff carved with the face of K'awiil, the patron god of kings.

The royal court, arrayed in jade beads and exquisitely woven cotton *mantas*, stood around the king, along with priests and royal wives, scribes, and offspring. Headdresses of all kinds perched atop their heads and above their tattooed faces and bodies, their filed teeth inlaid with green jade, pyrite or turquoise, their hands clutching wood and obsidian staves carved with the faces of the gods. Incense burners smoked, the aroma of the tree resin *copal* wafting in the air, carrying its scent skyward toward the Upperworld. Among them stood some of their vassals, including the king of Naranjo Sa'aal, Tajal Chaak, a cousin of Tikal's king.

Banners fluttered in the breeze atop the pyramid temple. Below on the plaza stood a crowd of thousands, interspersed among stalls with their own aromas of all kinds of foods and flowers. Everyone wore their best, the brightly woven cotton or clothing made from bark (fig tree or *Ficus* spp.) determined by who they were and their status—whether elders of well-known lineages, elites, merchants, *curanderos* (healers), ball players, or farmers.

The king raised his voice and called for rain. His plea carried from the temple, acoustically engineered so that performers at the top, royal or not, would only have to project their voices to be heard by the masses below. Calling upon his ancestors and the Rain God Chahk, he promised the first fruits of the harvest come August. Crops needed the life-giving rains. He called to his people to help him supplicate the gods and ancestors that have the power to bring an end to the annual drought. The king called upon his subjects to give to the gods food, objects, and even their own blood.

The sacrifice began. The head priest handed the king an intricately carved obsidian knife with a wooden handle inlaid with a mosaic of jade, iridescent shell, and turquoise. With the blade he perforated the tip of his penis. The dripping blood—the liquid of life/death/renewal—pooled into a bowl filled with bark paper, which quickly turned red. The blood-soaked papers were then lit. As they burned, the smoke rose to the gods above in sacrifice.

The king beheld a city of plastered and painted surfaces and towering pyramid temples bedecked with carved gods and other symbols erected by his ancestors now entombed within, and the crowning temples reflected in the waters of the reservoirs that lay amidst them. The kingdom of Yax Mutal had an abundance of fertile land but lacked lakes and rivers. Predecessors of Chak Tok Ich'aak II quarried out the massive reservoirs, and in the process provided building material for the increasingly grand temples and palaces. And it was his job to maintain the reservoirs. But they required rain. His subjects depended on him.

As king, Chak Tok Ich'aak II was responsible for making sure his subjects had clean water. With a long dry season, stagnant water in the reservoirs was a constant threat. Mosquitoes and parasites proliferate in such unclean water. And it was not drinkable. The king must keep his subjects and throne healthy and prosperous. Kings also had to work to maintain their hold over the tribute of others, especially when nearby kingdoms competed for labor using the same means to attract people—clean water. The kings' subjects drank, cooked, washed, made pottery and much else besides, notably the lime plaster that adorned temples and sealed plaza floors that funneled rain into the reservoirs.

After raising his arms from his moment of sacrifice, Chak Tok Ich'aak II looked down to the sea of faces, subjects who, while beholden to him, still could take away his power by withdrawing their support. He must perform well—to enthrall, to bring forth rain, fertility, bountiful crops, and prosperity. Chak Tok Ich'akk II knew he was born to rule Tikal, to implore its gods, to maintain its clean water sources, and to fight its enemies. He lived in a palace between reservoirs and temple pyramids, its rooms closed off from the common folk. His father, K'an Chitam, who ruled Tikal from 458 until c. 486 CE, performed the same rituals as he did, scion of a dynasty that boasted centuries of command over one of the greatest Maya cities. His mother, Lady Tzutz Nik, the daughter of Tzik'in Bahlam, the king of one of Tikal's subject cities, Naranjo Sa'aal, married his father to cement a political alliance. Both were buried in tombs within walking distance of the throne he sat upon in his palace.

Atop the temple, in the agony of performing his sacred duty, Chak Tok Ich'aak II's thoughts surely turned to the spectacle below him, the city, and the people. Below the king on the temple summit, friends and family living in different communities conversed with each other. Parents looked for suitable spouses for their daughters and sons. Women and men bartered food and items they made. All partook in the feast sponsored by the king and enjoyed the music and performances.

With the rain ceremony completed, other performances began near the ball court, such as reenactments of victorious battles, such as their 486 CE defeat of the nearby city of Maasal, and of the creation history, like what we find in the *Popol Vuh*, where the Hero Twins outsmarted and defeated the Underworld Gods and set their father the Maize God on the path to be reborn.

This glimpse into the life of one king played out similarly in other Maya cities for over 1,000 years. What brought it all to an end?

The Demise of Maya Kings

Maya kings kept and grew their power by attracting subjects with their clean water and by performing the essential ceremonies to keep the entire system working. Kings organized the building and maintenance of reservoirs for everyone to use in exchange for labor, goods, and services. In fact, the largest reservoirs are near palaces and temples. Royal life was at its busiest in the dry season during the seasonal lull in agriculture and when farmers needed drinking water since creeks, community *aguadas*, and many *bajos* dried up, just as they do presently. The more water they stored, the more subjects-slash-laborers kings could attract during the dry season. Rural Maya were the ones who constructed the urban reservoirs, temples, ball courts, palaces, and plazas. They were the ones who provided food and manufactured items such as ceramic vessels of all shapes, stone tools, grinding implements for maize, and other goods.

What brought down this political house of cards was the domino effect of royals diverting from the correct path and changing rainfall patterns, specifically, several prolonged droughts. Maya kings and cities had adapted to earlier droughts, such as the second century CE and early sixth century CE ones because there were fewer people, smaller cities, and fewer demands on the landscape. This adaptation stands in contrast to the ninth century severe droughts when kings continued to sponsor massive building and expansion programs that led to further land clearing and erosion in some areas. Had they diverted too far from the traditional inclusive, sustainable worldview? Perhaps. But they were excellent water managers—as long as there was enough rainfall.

An indicator of kings diverting from the "right" path is their use of the linear Long Count to document *their* history. Inscriptions focused on royal dynasties moving forward, not cyclically. Mesoamerican peoples used intersecting sacred and solar calendars that meshed on the same day once every 52 years. But only Maya kings used the Long Count (keeping linear track of time), whereas everyone else in Maya society used the solar and ritual calendars that intersected and focused on continuous cycles of renewal (the cyclical view of time). The zero date or creation date of the Long Count was August 13, 3114 BCE. The Maya are one of the few cultures in the world to invent the concept of zero. That said, they still included royal linear history within a cyclical existence—to a degree. The cycle begins again every 5125.25 years, or 13 *b'ak'tuns*. The current *b'ak'tun* began December 21, 2012, when it reset to zero.

The Maya experienced at least eight droughts 3 to 18 years in length between 800 and 930 CE. Droughts peaked in 806, 829, 842, 857, 895, 909, 921, and 935 CE, with the years between 804 and 938 CE showing a 36% to 56% drop in precipitation.[14] The multiple droughts explain why the political collapse took place over 100 years in the southern lowlands. Droughts had different impacts on the hundreds of cities, each with their own social, environmental, and political circumstances. They exacerbated any problems arising from erosion, deforestation, and/or overuse of resources. Common to all cities, though, is that reservoir levels plummeted, water quality worsened, water lilies died, and crops failed. People blamed those who had claimed such close connections to supernatural elements revolving around rain and fertility. But subjects did not resort to rebellion or violence. Instead, they deserted kings and cities in the southern lowlands *forever* by 900 CE. Most Maya dispersed in all directions, settling in smaller communities throughout the Maya area near permanent water sources, cities in the northern lowlands and highland Guatemala, and coastal areas. Royals in the southern lowlands once again became elite landowners as their forebears had been.

This urban diaspora was an extreme but necessary decision families made. They responded as anyone else would—to take care of their families. Farmers no longer continued to pay into a system that did not work for them. They voted with their feet and left the royal-urban fold. "History" became silent in the sense that inscriptions, for all intents and purposes, disappear in the southern lowlands. That said, farmers never had any need for the written word in the first place. The knowledge passed down through oral stories, histories, and experiences was all they needed. The Maya continued to take care of their fields, families, relationships, and their homes—just someplace else for about 90% of those who had formerly lived in cities and rural areas in the interior southern lowlands.

Maya kings existed for over 1,000 years—an amazing accomplishment. While kings have disappeared, their seats of power have not. Families are still here, too.

Interlude: A City Without a King

Even though water and royal power go hand in hand, there were Maya cities that didn't have kings, for example, several cities along the Belize River, including Saturday Creek (see Figure 7.2). Here, plentiful rich alluvium and the benefits of recession agriculture, higher rainfall than the interior lowlands, aquifers, and year-round water from the Belize River meant that there were no means for kings to attract subjects. There was nothing potential kings could provide that farmers could not acquire themselves; reservoirs were unnecessary. The only hierarchy was socioeconomic in nature; wealthier farmers had larger houses and could afford more exotic things.

Most of the city is situated in a modern cultivated field where farmers plow twice a year and plant maize, beans, watermelon, or black-eyed peas. Consequently, mounds have been reduced in height and spread in areal extent. The Maya lived here from at least 600 BCE through the early 1500s CE in farmsteads consisting of solitary mounds and mound groups or *plazuelas* (two to five buildings surrounding a plaza) dispersed around the city core, which includes a ball court, temples up to 10 m tall (31 ft), and elite residences. The city core has not been completely mapped because of dense secondary growth.

As mentioned in Chapter 7, Saturday Creek is a small city, built incrementally over time. Elites or wealthy landowners sponsored public events and organized the building of small temples, plazas, and ball courts for everyone's use. The Belize River provided plentiful clay for pottery, as well as a variety of fish, reeds, freshwater shells, and game that also relied on the river. And without "taxes" to pay, farmers could afford luxury items and exotics, including obsidian from highland Guatemala and central Mexico, jade from highland Guatemala, hematite and other materials, as well as beautifully painted ceramics. And it did not hurt that they lived along a major trade route. Further, the rich alluvium is excellently suited for crops such as cacao and cotton, which cannot flourish in most parts of the interior lowlands with their dark, well-drained soils better suited for maize, beans, squash, and other staples.

The political disintegration occurring in the interior southern lowlands in the ninth century CE did not have much of an impact on cities like Saturday Creek. Maya families weathered several droughts and continued living as they had done for generations. The Maya eventually abandoned Saturday Creek in the early 1500s CE—for as of yet unknown reasons. Perhaps the 1515–1516 epidemic, probably smallpox, that spread among the northern lowland Maya spread further south along trade routes. Or there was an earlier, unrecorded epidemic. After all, diseases spread faster than people. There is an unknown number of smaller cities with similar histories, for example, Aventura in northern Belize. Archaeologist Cynthia Robin and her team have been excavating Aventura for a decade and have revealed a place where Maya and their ancestors lived for nearly 5,000 years, from c. 3400 BCE through the early 1500s.[15]

***

Tropical political dynasties fell or transformed while farmers and former subjects endure. For the Maya, instead of diversifying their power base, kings put all their eggs in one basket, so to speak. Consequently, their power was vulnerable to any changes in rainfall. We know what happened. Most people emigrated out of the metropolitan heartland to new areas where different networks and sociopolitical and economic systems emerged. Urbanite political elites set up shop in more compact cities in different, peripheral areas, oftentimes with different political systems. In the end, farming families endured because of their reliance on small-scale water and agricultural systems which were more adaptable in the face of any drastic changes. In contrast, political powers depended on massive water systems that left them vulnerable and unable to adapt to any major disruptions.[16]

In other words, political strategies typically are based on short-term, unsustainable practices, whereas families and communities rely on long-term and enduring or sustainable ones.

Maya farmers existed before, during, and after the rise (c. 200 BCE) and fall (by 900 CE) of Maya kings in the southern lowlands. Political institutions come and go, but families endure, evidenced by the millions of Maya living in Central America and elsewhere. Maya kings, however, left their indubitable mark in cities, as I show in the next chapter.

Chapter 9

Yax Cities

Yax means blue-green in Yucatec Mayan. Maya do not distinguish these two essential and vibrant colors. It is the color of water and sky, symbolizing rain and fertility. It signifies the center of the universe. *Yax* also expresses a sense of "newness" or being "first." Here, *yax* conveys two key features of Maya cities: green (sustainable) and blue (reservoirs). They, too, cannot be disentangled.

We don't know the original names for most ancient Maya cities. We know Palenque's is Lakamha' or "Big Water." Tikal's is Yax Mutul or "First/Green-Blue Bundle." We don't know many of the names because some kings did not use hieroglyphs, archaeologists have yet to find them, or looters have stolen carved and inscribed *stelae*. Too many Maya cities have been breached by *huecheros* (looters), an apt term as it comes from how armadillos dig tunnels.

Maya cities had long histories, in some cases for over 2,000 years. The Maya built them to last. They included reservoirs and open spaces for agricultural fields, orchards with diverse trees, fishponds, and gardens. For instance, and as mentioned in Chapter 3, a few months after Hurricane Richard struck central Belize in October 2010, the Belize Institute of Archaeology asked me to assess hurricane damage to the Maya city of Yalbac (c. 300 BCE–900 CE). There was minimal damage, even though it hasn't been occupied or maintained for over a thousand years. The same goes for every other Maya city.

Most Maya walked away from southern lowland cities and kings over a thousand years ago. Their urban planning was sustainable—it was the kings who had failed. The *political* infrastructure fell apart, and people left. The fact that the empty cities have withstood hurricanes and tropical storms for over a thousand years without any maintenance speaks to their resilience and provides a roadmap for sustainable planning—if we can save them from destruction, urban sprawl, agriculture, and looting.

Up to this point, Maya cities have been in the background. It is now their turn for a closer look—especially how they lasted so long. I first cover the basics—how cities intersected with the rural world.

Cities and Farms: The Ties that Bind

Maya kings sat upon *pop* (woven mat) thrones. Their woven pattern signified an interwoven community. *Popol*, in fact, translates as "community." How did this system work when royals lived in cities and most subjects lived dispersed throughout the hinterlands? To begin answering this question, it is important to keep in mind the fact that farmers can survive without cities. But cities cannot survive without farmers and the local resource network that brings in forest and farm products, not to mention labor, something Bishop Diego de Landa noted in the sixteenth century in the Yucatán, Mexico:

> Beyond the house, all the town did their sowing for the nobles; they also cultivated them [the fields] and harvested what was necessary for him and his household. And when there was hunting or fishing, or when it was time to get their salt, they always gave the lord his share, since these things they always did as a community.[1]

Thus, it is not possible to understand cities without understanding the rural world.

Urban elites integrated water and agricultural systems, rural farmers, and exchange networks and resources, what is referred to as low-density urbanism, a concept first applied by Roland Fletcher to Angkor.[2] Farmers had political and familial obligations. They owed their king labor, goods, and services. They owed their families sustenance and security. When royal, superimposed demands lessened or disappeared, as they did by 900 CE when tribute demands faded away along with kingship, it made things much easier for families—to a certain degree.

Seasonality impacted sociopolitical systems: urban activities dominated during the dry season when cities drew in farmers with their reservoirs, public ceremonies and events, and markets. Rural living and agricultural activities dominated farmers' lives during the rainy season when they returned to their farmsteads spread throughout the landscape, mirroring the diverse but scattered fertile soils and other resources. The demands of urban royals and elites and a growing population ultimately resulted in a hierarchical political system to allocate resources, provide protection, resolve disputes, sponsor public events and building programs, and maintain the reservoirs.

This system peaked during the Late Classic (c. 600–800 CE) when population numbers and concomitant demands, such as land use and clearing, increased, resulting in decreasing evapotranspiration (i.e., the evaporation of water from soils and plants into the atmosphere) and a 5–15% decline in rainfall.[3] During the rainy season, rural farmers lived in self-organizing or heterarchical communities. The Maya farmed during this agriculture-intensive period and dealt with family and local affairs, including cooperative efforts to maintain the diverse and small-scale extensive and intensive subsistence systems. Urban occupants were wealthy landowners or elites, the royal family and their entourage, market vendors, religious and military personnel, administrators, and nearby farmers.

During the dry season the urban and rural worlds merged when rural Maya became enfolded into hierarchically based cities. The

Maya relied on urban reservoirs and conducted nonagricultural activities such as travel, construction projects, selling and buying in markets, and attending public events. They also contributed labor and services (craft specialists and hunters), produce (e.g., maize, beans, squash, manioc, cotton, cacao, pineapple, tobacco, chaya, and tomatoes), forest items (e.g., timber and fuel, construction materials, chert, game, fruit, and medicinal plants), and manufactured items (e.g., stone implements, ceramic vessels and figurines, and textiles).

While most households were self-sufficient to some degree, they did not make everything they needed. Some were better at making stone tools, for example. The more exposed chert outcrops there are, the less soil there is, providing opportunities to manufacture stone tools to exchange or barter for food in local or urban markets—as well as for tribute. The same goes for every type of household item and agricultural product. Not only did people exchange goods or foods at local and urban markets for things they could not supply themselves, but they also met with family, friends, and potential spouses for their marriageable offspring.

For access to dry-season urban reservoirs, everyone had to "pay" the "owners"—in this case, kings. Seasonal subjects built and maintained urban palaces, temples, ball courts, and reservoirs and replastered plaza and causeway surfaces frequently. Royals provided not only drinkable water, but closer access to the gods and divine protection. They did this through sponsoring and performing large-scale public ceremonies that brought everyone together to show their power and simultaneously their solidarity with everyone by conducting domestic and community rituals writ large—funerary, de-animation and animation rites, and rain ceremonies, as well as other ceremonies performed only by royals and their priests.

The lack of beasts of burden meant that human bearers transported staples and other items and foodstuffs. Maya bearers, using tumplines (a wide strap that rests on the forehead with two ends connected to a wooden frame), could not travel more than a few days without eating all the food they could carry. Maya also transported goods via canoes, which benefitted those living near waterways. But at the height of the rainy season, rivers are turbulent. At the height of the dry season, rivers are low and murky. Consequently, most staples were grown and consumed locally. Small prestige or exotic items such as cacao beans, jade, obsidian, feathers, and cotton mantas, were exchanged farther afield. The increasing number of *sak b'eh* (causeways) revealed by lidar suggests an expansive and fluid urban-rural engagement that likely extended into the rainy season. Such a system is illustrated par excellence at Caracol in Belize, where mapping and excavations show how the urban core and surrounding communities are connected via a complex system of *sak b'eh*, reservoirs, and markets.[4] I am sure that different sections of the *sak b'eh* are composed of different materials and/or construction techniques, reflecting the different communities who contributed materials and labor. Nonhumans used *sak b'eh* as well.

In sum, fertile land + farmers = plentiful food and surplus. Surplus funded royal lifestyles, building programs, and political infrastructure. On the one hand, this setup provided the means to produce enough food for family altars and shrines, local community temples, and royal temples. On the other hand, such a setup made it difficult for kings to reach anyone much beyond the cities over which they ruled, especially during the rainy season when agricultural needs were at their peak. Power, thus, was seasonal—during the dry season when farmers needed water. This system worked for over a thousand years.

The left side of Table 9.1 cannot function without the right side, but the right side can and did function without the left side. The left side "collapsed" by the end of the Terminal Classic period (c. 900 CE), and the right side continues today. The urban-rural symphony worked because of their inclusive worldview. Royals eventually broke this covenant and paid the price.

TABLE 9.1 Urban-rural interaction*

Urban	Rural
Centripetal (water, public events, and markets)	Centrifugal (diverse, dispersed resources)
Temples, palaces, large public spaces, orchards, gardens, and fields	Forests, fields, home gardens
Large-scale water systems (reservoirs and canals)	Dispersed, small-scale water systems (ponds, dams, and canals)
Hierarchical	Self-organizing
Royals, political and religious elites, merchants	Farmers and craft producers
Provides protection and potable water	Provides tribute (labor, services, and goods), i.e., surplus
Markets and production nodes	Household production and local markets
Inflexible, intricate systems—vulnerable	Diverse, flexible systems—resilient

* Adapted from Table 1 in L. J. Lucero. 2018. Climate Change and Water Management in Tropical Societies: The Classic Maya. In *Exploring Frameworks for Tropical Forest Conservation: Integrating Natural and Cultural Diversity for Sustainability, A Global Perspective*, edited by N. Sanz, D. Rommens, and J. Pulido Mata, pp. 204–213. UNESCO Mexico, Mexico City. Courtesy of the Creative Commons Attribution License (CC BY): https://creativecommons.org/licenses/by-sa/3.0/igo/deed.en

Seats of Power

In March 1992, archaeologist Anabel Ford and I flew over the Petén in a six-seater Cessna plane piloted by one of the best pilots in Belize, Javier. I watched with amazement—and a little trepidation—as Javier and Anabel used a National Geographic foldout map and a compass to figure out how to fly to the Maya cities we wanted to see from the air in Guatemala: Nakbe, El Mirador, Tikal, and Yaxha. At Tikal, the visible monumental architecture has been exposed by archaeologists (Figure 9.1). When you are walking around Tikal,

FIGURE 9.1 *Aerial photo of Tikal, 1992*

Source: Photo by the author.

you see the hundreds of other buildings covered by trees and centuries of soil and debris, waiting to be explored—and thus invisible from the air.

Every time I visit Tikal there is more deterioration. The ancestral Maya replastered buildings frequently, a requirement to counter the effects of the humid tropics. The current powers that be do not. Limestone dissociates in the face of direct sun—it basically becomes soft, crumbles, and turns into soil—that's why lots of the exposed buildings have turned black. In the past, all buildings not only were plastered white, but some were also painted with vibrant-colored designs in black, red, blue, yellow, or green.

In 1988 guerrilla activity had abated somewhat, so there were fewer military checkpoints on the main road to Tikal. At the time, there was no wooden staircase to help with the ascension to the summit of Temple IV. You had to climb the steep 70+ m (230+ ft)-tall temple using all four limbs, grabbing exposed roots and architecture and hanging on for dear life. The view is spectacular. You might be familiar with the view from Temple IV's summit—it was part of the rebel base in the movie *Star Wars Episode IV, New Hope*—thanks to George Lucas having his team carry up the film equipment.

Each Maya city has its own character and story to tell. Palenque, in Chiapas, Mexico, for example, with its stuccoed monumental buildings, four-story watch tower, and lack of *stelae*, is considered one of the most beautiful cities, especially with its mountainous setting. The Maya built systems to drain water away from Palenque because of its plethora of springs. Figure 9.2 shows the setting sun's rays in Palenque over the Temple of the Inscriptions, which contains the final resting place of its most powerful king, K'inich Janaab' Pakal, who lived to the ripe old age of 80 (603–683 CE) after having ruled for 68 years (615–683 CE).

FIGURE 9.2 *Palenque, 2005*

The Temple of the Inscriptions is to the left.

Source: Photo by the author.

One of the most amazing cities is Naranjo Sa'aal in Guatemala near the border with Belize. It is the size of Tikal, but mostly covered in jungle. It is truly magical. And sad. There's been so much looting. Thanks to the efforts of the project director, Guatemalan archaeologist Vilma Fialko, looting has ceased. But not before looters, with possible assistance from soldiers, removed *stelae* and hieroglyphic staircases from pyramid temples, looted tombs, and much more. More about looting at Naranjo Sa'aal in a bit.

***

The ancestral Maya would not be familiar with our modern notion of "cities." Maya cities did not have obvious administrative buildings per se or gridded streets or granaries. Instead, Maya urban layout reflects strategies royals used to draw people—and their tribute—in, that is, massive reservoir systems interspersed with orchards that provided fruit and shade, home gardens, fields, fishponds, patches of forest with flora and fauna, houses of all sizes, ball courts, palaces, temples, processional *sak b'eh*, markets, and plazas that could hold thousands of people.

Even though the most powerful kingdoms (i.e., Tikal and Naranjo Sa'aal in Guatemala, Calakmul in Mexico, and Caracol in Belize) largely arose in areas with fertile soils but without lakes or rivers in the interior southern lowlands, the Maya still built major cities along rivers, whose kings still relied on reservoirs to supply drinking water, especially at the height of the dry season when the rivers were low, murky, and disease-ridden. The larger cities supported 80,000 to 100,000 people (e.g., Caracol in Belize). To give you some perspective, during the same period as the Late Classic (600–800 CE), London had less than 15,000 inhabitants and Paris about 20,000.

Cities surrounded by less fertile land typically meant fewer farmers and, thus, fewer subjects and less power. There was never one king that ruled every city and all the dispersed subjects. Everything was just too dispersed to integrate, well, much of anything. The lack of major connecting road systems, and more significantly, the lack of beasts of burden to pull wheeled carts, as well as the lack of year-round navigable rivers, meant there were no means of transporting food, equipment, and troops long distances. That said, Tikal and Calakmul played the political game well enough to bring under their umbrella a changing sequence of cities through diplomacy, marriage, and battles among royal and elite warriors, particularly Calakmul.[5] Kings competed for subjects since they had choices where they went for dry-season water because the cities are only about 25 km (15.5 miles) apart from one another. Consequently, kings never had absolute control or rule over Maya farmers.

That said, inscriptions tell of kings with the elevated royal title *k'uhul ajaw*—divine or holy lord. A few kings had the power to use the even more exalted title of *kaloomte'*, which has been challenging for epigraphers to decipher. Many a royal history is inscribed on *stelae*, portable items like jade and marine shell and ceramic vessels, and staircases (e.g., Copán and Naranjo Sa'aal—until looters removed stair façades in about 2005) that record royal exploits, intercity interactions including marriages and ceremonies, dynastic successions, ball games, and ancestral histories. And not much else—except battles between cities.

Warfare between cities is a fascinating topic because battle dates and scenes appear somewhat frequently (e.g., Bonampak murals and Dos Pilas *stelae*), but material evidence of it in the archaeological record is less obvious or widespread.[6] Inscriptions mention battles, but only a few of the hundreds of cities show evidence of battles or violence, such as Aguateca and Dos Pilas in the Petexbatún area of Guatemala, Becan in Mexico, and a few others.[7] Without beasts of burden, warriors would need to carry their gear, weapons, food, and water. And since many areas are flooded becoming impassable or swampy during the rainy season, it would make sense to fight in the dry season. In fact, most inscriptions that mention battles or the capture of a king by another king happened in the dry season. Then the issue would have been finding water supplies to keep warriors hydrated.

We rarely find obvious war implements or caches of weaponry in the archaeological record; yes, we find spear points, but not enough in any given place to supply an armed force. And the Maya did not use arrow points throughout the lowlands until after 900 CE. For weapons, warriors likely used hardwood clubs, thorns sharp enough to puncture truck tires, poisons, spiked vines, and vessels filled with either hot coals or stinging insects like wasps, bees, and hornets.[8] These kinds of war implements would not leave evidence. Further, we don't find evidence for large-scale or extensive violence in human skeletal collections such as defensive wounds, broken and healed bones, or fatal blows. Violence appears to be at the individual level, as opposed to large groups—more in line with ritual killings, including human sacrifices and accused witches (see Chapter 4), murder victims, or executed criminals. Archaeologists have found embankments or defensive walls or palisades, but not enough to warrant there being large-scale "Maya warfare" per se. And while lidar has revealed a noticeable number of hilltop constructions some archaeologists label as fortifications,[9] why would warriors leave nearby cities open to attack if they fought on hilltops? Excavations will show if they functioned as fortifications—or more likely, as temples or shrines as part of *wits* (lineage mountains).

There wasn't a pan-Maya war god comparable to K'inich Ajaw the Sun God, Chahk the Rain God, or the Maize God. Societies where warfare plays an important part in politics have war gods: the Aztecs had Huitzilopochtli, Rome had Mars, and Greece had Ares. In Karen Bassie-Sweet's excellent book, *Maya Gods of War*,[10] she talks about aspects of Chahk, K'inich Ajaw, and the central Mexican Storm God Tlaloc when talking about war—but not about an actual pan-Maya god of war.

Dispersed subjects would have been hard to conscript as warriors or soldiers—not to mention the fact that farmers-slash-subjects had choices which king to provide their labor, goods, and services to in exchange for access to dry-season water. Inscriptions tell of "battles" or "axe" events and the like—but they likely involved a small group of elite or noble warriors and the king and his entourage and not much more than that. We always need to take the written word, especially those written by the victor, with a grain of salt.

One place where intercity battles may have taken place was ball courts. Ball courts, like many other things in Maya life, had multiple purposes. The Maya played ball games in them and re-enacted the parts of their origin history that took place in a ball court in the Underworld. Ball courts often serve as entrances to the main royal plaza like we see at Tikal and other powerful cities. They would have also served as an excellent means to handle disputes—where teams from competing cities played one another.[11] The losing ball players from the "defeated" city may have been killed, as depicted on the sidewalls of the massive ball court (96.5 × 30 m, or 316 × 98 feet) at Chichén Itzá in the northern lowlands. In an example in the southern lowlands, the carved limestone La Corona Ballplayer Panel 1 depicts a ball game that took place in 687 CE at Calakmul, Mexico, played between Calakmul's king, Yuknoom Yich'aak K'ahk' and a noble or priest from La Corona in Guatemala, a subordinate city.[12]

Whatever the purpose and scale of "Maya warfare," southern lowland kings existed for over a millennium. Ironically, the more kings depended on reservoirs to attract subjects, the more vulnerable they became to any changes in rainfall patterns. Kings did not diversify their source of power, and they should have. In sum, kings were vulnerable, while cities were resilient.

Still a "Mystery"

I typically avoid the term "mystery" when discussing the Maya, but sometimes we really are dealing with a mystery, highlighting how much there is still to learn about the ancestral Maya. And we are in a race against time, looting, urban sprawl, forest clearing, plowing, and outright destruction (e.g., ancient Maya buildings bulldozed and used for road fill).

The first mystery has to do with the ancient Maya city of San Jose in central Belize north of Yalbac. This city is famous in Maya archaeology. Sir J. Eric Thompson excavated it in the 1930s.[13] Because of the many construction events his team exposed while excavating temple pyramids, Thompson was able to devise a fine-tuned ceramic chronology based on changing ceramic styles that is still used today to date sites. This was before radiocarbon dating—which has since supported the time periods ceramic specialists generated. As for San Jose, no one has been able to relocate it other than in 2008, after which it was "lost" again (we were unable to get GPS readings due to the dense jungle canopy). You must remember that old maps, such as Thompson's, showing the location of ancient Maya cities in the middle of the jungle are just approximate. San Jose bears this out.

The jungle hides much and has lots of secrets. As of the writing of this book, San Jose still has not been relocated, though Cleofo has informed me he is sure he knows how to find it. I believe him.

I have one final "mystery" to share. If you were to search on Google Earth for Maya sites west of Yalbac toward the Guatemalan border on the property now owned by the Belize Maya Forest Trust, the area near the border appears to be devoid of sites. But it is not. It is just that this area has not been explored by archaeologists. While conducting tree surveys, Yalbac Ranch loggers came upon two unnamed and unexplored ancient Maya cities near the Guatemalan border that I briefly visited in 2005.

One of the cities (UEC1) is less than a kilometer from the border and is about the size of Yalbac. It has an acropolis, temples, and a ball court. The other city (UEC2), a tad northeast of UEC1, is larger. While this second city has temples, an acropolis, and large plazas, it also has a massive wall/structure on the edge of what appears to be an artificially flattened hilltop. And, of course, we saw lots of substantial looters' trenches bisecting temples and other monumental edifices.

In 2018 Larmon, Stanley Choc, and Tilo Luna, with assistance from Yalbac Ranch supervisor Esteban Alvarez, attempted to relocate the cities to gather enough information to apply for funding to map and excavate them. They still had to deal with the repercussions of the 2010 hurricane—dense, hot, and sticky secondary undergrowth. The exploration team went as far as they could on a skidder (a heavy vehicle that pulls logs), after which they walked several miles and arrived at UEC2. Unfortunately, since they still had to get to UEC1 and set up their hammocks before nightfall, they only had a brief time to quickly draw a map of the city—and all the looters' trenches. They didn't have time to relocate UEC1, even though they walked for four hours the first day and another four the second day. They eventually had to turn around, as they were running out of time and, more importantly, water. They, along with one of Cleofo's sons, Mark, tried and failed again to reach UEC1 in 2022. It was the end of the dry season, and the landscape was particularly parched. They could not carry enough water to last the long trek by foot working their way through the jungle with machetes. However, in the process of trying to reach UEC1, they came upon another city (UEC3). Today, it is difficult to imagine that there is still so much more to explore. These kinds of stories highlight how well the ancestral Maya adapted to similar conditions for millennia.

These stories also go to show how much more densely populated this area was in ancient Maya times—yet there were still forests, clean water, fertile soils, and a plethora of flora and fauna. The Classic Maya fed way more people in the past than are fed presently, and without massive deforestation. The biggest danger to these hidden sites, rural and urban, is looting.

Looting

There are several threats to archeological sites worldwide. One is encroachment—by human settlements and resource extraction, especially mining and logging. More recently, climate change has been a growing concern to ancient sites globally—rising sea levels, frozen and protected sites being exposed due to warming weather, and increasing wildfires. Like what happened at Barton Ramie, where farmers bulldozed ancient Maya monumental architecture to use as road fill, this destruction happens too frequently with minimal repercussions. There are too many examples to illustrate this point.

A major problem is looting. Looting has been happening in the Maya area since the first explorers—all you have to do is visit major art museums. When you see, for example, a beautifully painted Maya cylindrical vase and the description reads something like, "Classic period polychrome Maya vase, Guatemala," you know it is a looted piece because it lacks detailed provenience information, that is, from which city, building, unit, and stratigraphic context it came. Looters destroy and desecrate ancient Maya tombs and homes to plunder them for vibrantly painted ceramic vessels, carved pendants, earrings and masks made of jade and shell, obsidian ceremonial knives and objects, and carved and painted and inscribed sculptures and wall panels.

The repercussions of looting are many. We lose so much information. The context of where an object comes is sometimes more telling than the object itself. Let me illustrate with a Mont Blanc fountain pen. Let's say the president of the United States used such a pen to sign major legislation into law. Without context, it is just a beautiful Mont Blanc pen. With context, the story has rendered the pen priceless and part of history. In brief, we lose lots of history because of looting, and the Maya lose their cultural heritage. This cultural heritage is vital not only for descendent communities, but for what lessons or ideas it contains.

The jungle hides ancient Maya sites and makes it easier for looters to wreak havoc because it hides them, too. Nearly every site in the jungle has evidence of looting, including the unnamed and unexplored cities. Also, bats love looters' tunnels that breach temples and other monumental buildings, which means lots of *guano* (bat dung), which can be dangerous because it can cause respiratory problems (e.g., histoplasmosis). It is critical in view of this reality that we learn as much as possible about this amazing society before evidence of its existence disappears—not just because the ancestral Maya built a great society, but because there are lessons to be learned that are relevant in dealing with current issues regarding global climate change and devising sustainable practices.

Who are the looters in the Maya world? It could be anyone—Maya, non-Maya, or foreigners. The worst are the dealers who know what they are buying is illegal, pay little, and resell at exorbitant prices (and profits). One of the worst instances of looting is at the ancient Maya metropolis of Naranjo Sa'aal in Guatemala near the Belize border, which I first visited in 1989 and mentioned briefly earlier. It is currently under consideration as a UNESCO World Heritage Site.[14] I next went in 2007 with Vilma Fialko, the director of the Naranjo-Yaxha archaeology project. She was forcibly kept out of Naranjo Sa'aal for three years by looters. The owner of a major Guatemalan company, possibly assisted by soldiers, paid looters to loot, and did they ever. Vilma once even found a headstone with her name, face, and date of death inscribed on it. Destroyed *stelae* lay strewn about; looters sawed off a thin slice of the carved surface and left the rest behind. The number of looters' trenches (270) at pyramid temples and palace tombs is staggering. The stripped hieroglyphic staircases are distressing to see as well. And it is such a beautiful, magical place. Some looted items have been returned to the government, some are in museums in Guatemala, and some are still in private hands. But the damage has been done, not only to the ancient city, but to Maya cultural heritage.

Vilma and her team recorded the 270 looters' trenches. They drew wall profiles, took photos and measurements, and collected exposed artifacts. Once recorded, they backfilled the trenches and then planted different kinds of flora to identify the location of the trenches. These plants also help maintain architectural integrity. Her team uses different plant species to distinguish excavations and other types of explorations. They also planted begonias as ground cover to protect soils and buildings. They are recreating a garden city. Currently, over 50 endangered flora species are flourishing at Naranjo Sa'aal; they seem to like the steep sides of monumental structures, and botanists consider this ancient city to be a biodiverse wonder.

The main reason we need to save the past is to benefit our future: there is so much ancestors can teach us in moving forward. In doing so, we would also save people's cultural heritage. Looters begone, so we can learn, implement, and survive.

Chapter 10

The Survival of Our Planet

The sixteenth century K'iche' Maya origin history, the *Popol Vuh*, tells us that currently we are living in the fourth world. The Creator Gods destroyed the earlier ones because they were displeased with the humans they had created. Will the Gods destroy this world? Or will we do this on our own via the overuse of nonrenewable resources and deforestation and the human-induced climate change caused by our addiction to greenhouse gas–producing fossil fuels and livestock?

It is possible to change the course of our and the planet's history, but it will be challenging and hard work. Maya wisdom and insights show us a way, and not just for the tropics, but globally. Despite the last five centuries of conflict, forced migration, and nucleation; exposure to foreign diseases; disenfranchisement; and droughts, Maya resilience comes from how they lived and engaged with their world. The over 7 million Maya living today are a testament to this fact.

What have we learned from the Maya about how to survive over the long term? First, the ancestral Maya acknowledged nonhuman kin and family connections. And family takes care of each other. The Maya coexisted with nonhuman entities or beings. Neither over-taxed the other. We know this because the ancestral Maya did not cause the extinction of any plants or animals. This reminds me of sharks. Sharks have survived largely unchanged for hundreds of millions of years without killing off or depleting the "resources" upon which they rely—prey. *Homo sapiens*, however, have been around for over 200,000 years and are depleting the Earth's resources—and have left a massive footprint and lots of scars in the process.

The solutions I present are inspired by Maya insights I've learned from 35 years of working with the living Maya and learning about the ancestral Maya as an archaeologist. The solutions are relevant for all families or households, however defined or structured, because action and change start in the home and continue in communities, cities, governments, and transnational corporations. It will take all of us.

Insights and Solutions

I reiterate key points from previous chapters to highlight relevant insights and the fact that "For the Maya, human society is intimately involved with plants and animals.... There is no opposition of 'man' and 'nature'; the world is a garden."[1]

Diversity

Diversity is key, just as it is for exercise regimens, vitamins and nutrition, genetics, and financial portfolios. It spreads risk. Ancestral Maya families did not put all their eggs in one basket. They planted a plethora of plants and trees in their home gardens and *milpas*. Diverse agricultural practices (e.g., terraces, dams, canals, and raised and drained fields) and forest collaboration sustained them and the forest for millennia. Traditional knowledge passed down for millennia shaped their diverse, reciprocal relations with the tropical forest, agriculture, weather prediction, fauna, and more.

We, however, have put all our eggs in one basket, that is, technology fueled by fossil fuels.

Solutions

Diversify everything—multicropping, heterogeneous gardens, novel energy sources and truly green technologies, and other strategies that take nonhumans into consideration. The remainder of this chapter is replete with examples of diverse strategies we can adopt (e.g., edible and medicinal plants in gardens, empty lots, and other based open areas). We also need more diverse technological innovations based on renewable energy and innovative green technology that does not rely on finite resources to construct and/or fuel to keep running. For instance, current green technology relies on batteries (e.g., electric vehicles) manufactured with finite amounts of minerals that must be extracted from the Earth (e.g., lithium, nickel, cobalt, graphite, manganese, alumina, tin, tantalum, magnesium, and vanadium), leaving scars on the land and local peoples.

Inclusive Worldview

The Maya worldview does not privilege humans. Unlike in current Western thinking, the ancestral Maya did not separate culture and nature but, rather, were on the same plane as animals, land, water, soils, things, and other entities. Everything was animated and had a *ch'ulel* (soul). They all coexisted and communicated with each other through their souls. Everything was and is connected.

Today the Maya continue to engage with the tropical world with remnants of their worldview via their traditional knowledge of forest collaboration, weather and seasonal patterns, and agricultural practices. They also continue to conduct ceremonies to communicate with ancestors and gods, which now include Christian ones, and for purposes of transformation and renewal. Their worldview embodies diverse interactions and overlapping and interwoven relations, all for the same purpose—world maintenance. The moral of the story: it takes everyone—humans and nonhumans alike—to ensure *our* continuance.

Many preindustrial or non-Western traditional societies had or have worldviews similar to that of the Maya, the common theme being the concept of family or kinship that includes nonhumans with mutual responsibilities to each other and their place of existence.

Solutions

The family or household is at the core of sustainable living. Each household is a small but essential part that together can start a revolution. We can turn small-scale responsibilities and actions into movements.

We can expand our notion of family and kinship—and responsibilities. Just look around you. We rely on so many entities. What are the things that you cannot live without made of? Where do they come from? Are they local? Imported? Do you ever acknowledge your reliance on fellow entities? Or do you take them for granted? We talk to plants and animals all the time, so why not other entities? Many of the things we buy make our life easier. It is convenient. Convenience will be the death of us and Earth if we don't step back and reevaluate its costs in terms of, for example, deforestation, greenhouse gases, depleted resources, and pollution. Further, the lack of engagement with nonlocal goods and services means that we don't see the impacts and ramifications on things produced thousands of miles away.

Traditional knowledge is essential in developing plans of action, as is avoiding path-dependent and inflexible strategies that are vulnerable to any disruptions. This knowledge, from Indigenous peoples who have been living in any given area for millennia to non-Indigenous multigeneration families or small stakeholders, is critical for any future planning because experience matters. Implementing plans without their input may work in the short term but ultimately fail in the long run.

For examples of the perils of ignoring Indigenous knowledge, read *Jungle: How Tropical Forests Shaped the World—and Us* (2021) by Patrick Roberts, where he tells the sad—and avoidable—story of the detrimental impacts of colonial rule and economic policies devoid of input from Indigenous peoples, which continues to have negative repercussions to this day. Colonial policies are an example of top-down mismanagement at its worst. Things are starting to turn around to a small degree, but more needs to be done. For instance, the National Science Foundation's Science and Technology Centers Program is funding the Center for Braiding Indigenous Knowledges and Science (CBIKS) at the University of Massachusetts at Amherst, which "will examine how to effectively and ethically braid Western and Indigenous science research, education, and practice related to the urgent and interconnected challenges of climate change, cultural places, and food security."[2]

Terminal Classic Maya kings also were guilty of making top-down mistakes, like when they used essential resources to sponsor increasingly ornate ceremonies and building programs between 800 and 900 CE instead of exploring other avenues in the face of several severe droughts. They failed to recognize or acknowledge that their previously tried and true source of royal power, reservoirs, was drying up.

The colonial mantle, in the sense of top-down management determined by profit margins, is now also worn by transnational corporations, which are run by people with families. We must work together—it is not enough to get the business world to adopt greener practices. They, too, must completely transform how they think, even if it means completely changing their line of business.

Transformation and Ritual

The Maya live their inclusive worldview daily, expressed through rituals. Family ceremonies solidify people's membership in the household, family, society, and the cosmos. They have performed the same rites for thousands of years in their homes and home gardens, as well as in the forest, *milpas*, water bodies, and openings in the Earth or *ch'e'n* (portals) to the Underworld. The same rituals were later also performed by kings on temple summits or in public plazas to promote solidarity and integrate more subjects. Rituals acknowledge reciprocal relations with other people, fellow guardians of the Earth or nonhuman entities, and ancestors and gods such as the Maize God, K'inich Ajaw the Sun God, Chahk the Rain God, and others.

Maya invested in place for millennia, materially and spiritually, to acknowledge their connection to others. They integrated and assimilated into existing systems rather than differentiating themselves, expressed par excellence in their homes. What better way to establish long-term relations than to inter deceased family members within their homes? The dead thus played a role in the lives of the living as ancestors. Commoners buried family members about every *katun* (20 years) when thatch roofs needed replacing. Death, de-animation, renewal—again and again. The history of building, destruction, and rebuilding is a history of transformation and renewal. Everything is renewed and balance is maintained. The Maya transformed other entities to survive and, in the process, changed relations, but without detrimental consequences.

Rather than separating the landscape into cultural and natural categories, ancestral Maya merged their lives with it. Once permission was granted by the Earth God to use his materials—mud, palm leaves, branches, stone, logs, vines as twine, and more—the Maya transformed one animated form into another through offerings, prayers, and the burning of incense.

Maya rituals have changed relatively little over the millennia. The number of ceremonial offerings in the archaeological record is staggering. Even with all the production and consumption of ritual objects, the ancestral Maya did not overuse resources; for example, despite the millions of Maya ceramics, we have not yet found a ceramic production site because it left a minimal footprint—it was a type of sustainable transformation.

This way of life (and death) differs from ours, where individuality and mobility are the norm.

Solutions

We can share and spread transformative ideas via familiar or traditional rituals. Ritual pervades all aspects of life. Knocking on wood; throwing salt over your shoulder; celebrating weddings, funerals, and birthdays; praying, going to religious services; and many more. Most of our current rituals focus on the individual, though. We can expand rituals to include others, human and nonhuman.

Rituals have multiple meanings. Rituals integrate. They promote belonging, unify people from all walks of life, and induce change within traditional cultural constructs. Rituals don't change much, but the beliefs surrounding them do, making them an ideal way to insert change. Rituals promote solidarity, more specifically, "solidarity without consensus."[3] Solidarity is created by people participating together—not by people thinking or believing the same thing per se. Let me illustrate. Among the Weaverland Conference Old World Mennonites of New York and Pennsylvania, when asked about the significance of the "kiss of peace" during church service, everyone has a different response: "Our ancestors brought the kiss with him or her from Germany and Switzerland," or "The kiss was established by the early leaders of the church," or "It's from the Bible. The disciples did it," or "I don't know why we do it."[4] One just has to "make it look right." Basically, as long as congregants follow the rules, personal beliefs are of little consequence.

This is why ritual can be a powerful mechanism for change. Rituals bring people with different mindsets together. Take politics, for example. Leaders throughout space and time have inserted change via traditional ceremonies conducted in larger and more public settings to integrate more people and introduce different political ideologies. New ideas and customs are introduced—and adopted. We can do the same but include nonhumans/fellow guardians to emphasize our connections and relations. For the ancestral Maya, precursors to kings performed household ceremonies writ large in plazas where thousands of people participated. But all Maya still conducted the same rituals—ancestor veneration and animation and de-animation rites—in their homes, further promoting solidarity because everyone, from farmers to kings, performed the *same* rituals.

Rituals integrate top-down and bottom-up movements. For the former, this includes government, transnational corporations, and the major world religions. For the latter, this includes families, community organizations, churches, and other social groups. Let me show you what happens when *only* top-down approaches are implemented. In the 1970s the government of Bali implemented modern agricultural strategies as part of the Green Revolution with its extensive monocropping, plowing, and use of chemical pesticides, herbicides, and fertilizers. It was a dismal failure. Government officials then proceeded to ask religious leaders and anthropologists how to revert to the sustainable, traditional, dispersed and diverse agricultural strategies that revolved around a ritual calendar and temple priests via temple districts that emphasized rotations and noncontiguous fields.[5]

One idea to promote solidarity and new ideas comes from former President Obama's successful grassroots campaign strategy for the 2008 presential election.[6] For his first presidential campaign, "By taking to heart the mantra of the field campaign, 'respect, empower, include,' a small group of paid and unpaid organizers went out into the streets and the suburbs and started a movement powerful enough to overcome... attack ads, robocalls, and smear tactics." They would not give up. Social media played a large role via neighborhood Facebook and Instagram pages, as well as door-to-door campaigning. These grassroots efforts resulted in the election of the first African American U.S. president. A key factor was recruiting young people, a strategy that resonates today—it is their future we need to save.

Healthy and Biodiverse Forests

The Maya interact with the forest like a garden—and with the garden like a forest. The Maya see more, so much more, when they peruse "nature" around them—comparable to a pharmacy, a food market, a hardware store, and a general store. Traditional Maya live as part of an ecosystem, not separate from it. The Maya rely on and have protected forest biodiversity for millennia and mimic its diversity in their gardens and fields as I've shown with the Choc home garden (see Figure 1.1), which includes flowers, edible and medicinal plants, poisons, and trees to use in construction and to make household items.

Healthy and biodiverse forests are indicated by long occupation histories. We see this at the small city of Saturday Creek, where Maya families lived for over 2,100 years, which they could only have accomplished through maintaining and collaborating with local resources/entities. In fact, the ancient Maya landscape was a mosaic of green cities, rural farmsteads, *milpas* and fallow *milpas* that still provided food and attracted game, forests, *bajos*, and untouched sacred places like Cara Blanca.

The Maya have diverse direct and indirect reciprocal relations with the tropical world—direct via collaborating with the forest and indirect via pilgrimage. The forest provides a plethora of food and items for daily use—fruits, nuts, roots, and game; natural fibers for mats, clothing, rope, twine, and hammocks; hardwoods for posts and vines to latch them; clay; minerals and organics for paints; chert outcrops the Maya quarried to make stone tools; and tree resins that are either burned as incense for offerings or made into the rubber balls used in ancient Maya ball games in the ball courts. Additionally, forest items typically have multiple uses—palm trees supply palm heart, oil, and leaves for shade, waterproof thatch roofs, and brooms. Limestone is turned into construction blocks and plaster floors, is used to soak and cook maize, is powdered to kill insects, and is crushed to use as temper in pottery manufacturing. Allspice's dried berries are ground and used as an aromatic spice and as a medicine for gastrointestinal ailments. Its leaves repel insects. There are so many more examples.

Another type of collaboration is how the Maya engaged with pilgrimage destinations and presumably other "sacred" places.

Pilgrimage takes place via a ceremonial circuit that connects people and places and renews relations with the forest and its denizens, as we saw at Cara Blanca, Belize. Such areas with minimal or manageable human occupation and construction promote conservation, biodiversity, and gene banks.

Collaborating with the forest, soils, water, and nonhumans in diverse ways encouraged subsistence flexibility and community investment in their home, making the entire ecosystem resilient. Maya relations with the forest worked, and widespread deforestation didn't happen. The Maya maintained the forest, and the forest maintained the Maya.

Just a reminder: there is no Mayan word for "weed," just an "un-useful" plant.

Solutions

We can collaborate with forests. We can reforest areas with native trees and plants. Forests play a vital role on Earth, and trees produce oxygen, absorb carbon dioxide (CO_2), and support life. They provide shade and are home to innumerable species. Deforestation is the opposite and increases temperatures and humidity, decreases rainfall, increases aquatic sedimentation, decreases water quality, and exacerbates the spread of pests. And while you might expect that more diverse plants would lead to an insect explosion, this is not often the case, and when they do occur, they are typically due to human disturbance and actions (e.g., only planting one kind of crop and overuse of insecticides).[7]

Logging companies can log more sustainably. With cedar, for example, "The wood itself is dead supportive tissue, so the harvest of a few boards from a big tree does not risk killing the whole organism—a practice that redefines our notions of forestry: lumber produced without killing a tree."[8] We can thus acquire timber without killing trees. Minimally, we can replace the trees we harvest. Yalbac Ranch loggers never clearcut in over 30 years of logging, but instead culled from the jungle in central Belize. And they planted four saplings for every tree they cut down.

We have endangered and caused the extinction of many fellow entities. We can cease hunting and fishing for sport—only kill what we eat and for conservation purposes. For fishing, minimally make capture-and-release the norm. Overfishing and overhunting disrupt ecosystems and interrupt the cycle of life. Thus, it is necessary for us to restore balance through, for example, sanctioned hunts that are not conducted just to protect humans and their economic concerns; that is, we need to include nonhumans or other entities when deciding to cull. Consider, for example, recent evidence that shows the detrimental impacts of nonsustainable hunting. Using International Union for the Conservation of Nature data on about 47,000 species, Chris Darimont and colleagues demonstrate that people hunt, fish, and collect kill or capture about 15,000 different species, or about 33% of vertebrates on Earth, and compared to nonhumans, exploit 300 times the number of species.[9] Adding to these high numbers is human demand for medicine, food, and pets. Consequently, about 40% of exploited species are threatened. These trends have and will continue to have detrimental consequences for biodiversity and ecosystem health. We can reverse this trend by preserving and promoting forest biodiversity, providing gene banks, and supplying potential new sources of food, medicine, and construction materials. Greater plant biodiversity also promotes soil regeneration.

We can create and expand local, regional, and national parks that that will not only showcase their beauty, but also conserve flora and fauna. We also need to rethink "parks" and include input from Indigenous peoples, as well as inviting them to return as stewards of their homelands. After all, they successfully and sustainably collaborated with forests for millennia in many parts of the world. For example, in the United States, President Biden supported a November 2021 joint secretarial order (Secretaries of the Interior and Agriculture) that resulted in over 20 costewardship agreements with different tribes (e.g., fire management and prevention). There are an additional 60 costewardship agreements in process that involve 45 tribes.[10] There is so much more we can learn from stewards from around the globe. For instance, "Our truth...in a majority of Indigenous societies, conceives that we (humans) are made from the land; our flesh is literally an extension of the soil."[11] Would you desecrate family members? Or would you instead visit, negotiate, acknowledge, and collaborate with them?

Water

It is life. Clean water, especially. In the Maya world, everything is rainfall dependent. Predicting when the rains will begin is challenging. If farmers plant too soon, the seeds will rot or ants will eat out the soft parts. If they plant too late, the seeds will not germinate. Maya continue to perform Cha'a Cháak (rain) ceremonies to propitiate Chahk for the ideal amount of rain at the right time.

The ancestral Maya adapted to unpredictable seasonality by constructing self-cleaning reservoirs, more specifically, constructed wetlands (CWs) that supplied millions of people throughout the dry season. Maya urban planning basically revolved around water diversion, capture, and storage. The city *was* a water management system.

Solutions

In the United States, people voluntarily could turn the over 10,000,000 residential and over 300,000 community swimming pools[12] into constructed wetlands (CWs) using aquatic plants and other organisms.[13] They would provide clean drinking water, as well as fish, snails, shellfish, fertilizer, edible and medicinal plants, and reeds for basketry. And you could still swim, of course. Companies already are producing natural swimming pools (e.g., BioNova Natural Pools). For natural ponds and the like, there also is the ScumSucker® aquatic weed puller that does not use herbicides but extracts aquatic plants by the roots, after which excess water is removed and the plants are compacted to use as, for example, fertilizer.[14]

CWs prevent standing water from turning into cesspools or breeding grounds for water-borne diseases and pests. This natural treatment system is better than purifying water with chemicals as is done currently in water treatment plants. In fact, civil engineers recognize the potential of CWs to manage wastewater and are trying to find wider applications of self-cleaning water bodies (e.g., Sedlak Research Group, University of California, Berkeley). CWs do not need chemicals or fossil fuels to run, and they become self-sufficient with some maintenance after the initial labor-intensive output.

Using more CWs also would contribute to fulfilling the United Nations Sustainable Development Goal 6[15] to ensure access to clean water for everyone, as well as encourage communities to work together to improve local water quality. People can work together to provide their communities with clean water starting with small CWs that they could also plant trees around like the Maya did. CWs are being used throughout the world, including Europe, China, North America, India, Sri Lanka, and Bangladesh. I am sure ancient Maya reservoirs, most of which currently have no water due to the lack of maintenance for the past 1,100+ years, can be rejuvenated as CWs.

We also need to conserve water and other entities. I learned in Belize that "if it's brown, flush it down, if it's yellow, things are mellow." Take short showers, which includes turning the water on and off while shampooing and soaping up. There's lots more we can do, like turning lawns, which require lots of water, into biodiverse gardens with native vegetation that supports birds and pollinators and other fauna and with shade and fruit trees. We can collect rainwater like the *tinacos* (water tanks) seen throughout Mexico City on the roofs of buildings of all types. Finally, we need to protect current wetlands and other water bodies worldwide, especially since we have lost over 80% of natural wetlands since 1700 CE.[16]

Agriculture and Subsistence

Maya reliance on the diverse domesticated and nondomesticated foods they planted or collected from the forest served them well in the face of seasonal changes, extreme or not. Planting diverse crops in a variety of places such as *milpas* and home gardens spreads risk. As mentioned previously, the Maya use over 500 native plants. And since everything blossoms at different times, fresh "produce" is always available. Blurred boundaries exist between forests, gardens, and *milpas*.

Traditional Maya life revolves around maize, beans, and squash—the "Three Sisters." No monocropping here. At present, monocrops are at risk of fungal diseases including wheat stem rust, rice blast fungus, corn smut, potato late blight disease, and soybean rust, all of which are becoming more and more resistant to fungicides.[17] In contrast, the Three Sisters are like a well-composed symphony, one that extends to other crops. Plant diversity translates as insect diversity—including nurturing ones that consume fellow insects ("pests"). The Maya also plant noncontiguous plots to prevent the spread of disease. Also preventing the spread of disease is their use of a plethora of plants, since each has different qualities—some are less susceptible to disease, drought, insects, or other banes of agricultural crops.

Ancestral Maya farmers also relied on diverse subsistence strategies including raised and drained fields, dams, canals, ditches, and terraces in home gardens; short- and long-fallow fields near and far from home; and combinations thereof. Fallow *milpas* attract game and continue to provide materials to build structures and produce like fruits, nuts, root and tuber crops, and botanical medicines.

Solutions

We can implement different modes of agriculture, from family-owned farms to large-scale agribusiness ones. Jane Mt. Pleasant, an agricultural soil scientist of Haudenosaunee (Iroquois) heritage, shows that Indigenous farmers in the Americas relied on intensive and productive farming strategies, which was much more common than the assumed widespread extensive slash-and-burn method.[18] Farmers used the latter in areas with poor soils or as the first step in creating permanent fields. Further, she found that farming on fertile soils without the use of plows has a much less negative impact than relying on plows and draft animals; tilling is more sustainable than plowing. Over time, plowing depletes soil organic matter through oxidation, which results in the loss of nitrogen and phosphorous. There is much less oxidation in fields without plowing. Mt. Pleasant and her colleagues have found that the best way to replenish nitrogen is through planting legumes—alfalfa especially, as well as clover, which helps to rebuild soil organic matter. In Belize, Maya farmers plant pumpkin, okra, and other nutrient-rich crops to replenish soils in between periods when they grow their main crops.

There are other, more sustainable alternatives for agriculture as well, including sustainable intensification and working lands conservation. The goal of sustainable intensification is to redesign current land use via both ecological (e.g., diversification and nonchemical pest management) and technological means without cultivating more land and further damaging the environment.[19] Working lands conservation attempts to lessen technology-intensive agriculture and highlight traditional knowledge and diverse management—or, rather, collaborative—strategies to promote sustainable food production and increase yields, such as agroforestry, forest collaboration, silvopasture, and diversified farming.[20] Sustainable intensification and working lands conservation also benefit the environment by supporting biodiversity, pollinators, and wildlife corridors. Both methods highlight diverse yet productive strategies. And we need to incorporate traditional knowledge to design enduring or sustainable practices that are healthy for humans and nonhumans alike.

We need to rethink residential and commercial landscaping. We can replace lawns with diverse native trees and vegetation that can include edible and medicinal plants. Labor input is relatively low to plant a diversity of tree species and create orchards over the long term. We can plant noncontiguous plots to prevent the spread of pests and diseases. People can plant trees whose decomposing leaves increase soil productivity, like the large leaves of the cohune palm (*Attalea cohune*) in the Maya area. The ancestral Maya composted, too, using nonhuman waste such as ash and food scraps, which we know about based on phosphorus soil analysis from areas inside and outside Maya residences.[21] Engaging with soils not only makes them more fertile, but can capture carbon in noticeable amounts, as seen in the Amazon.[22]

We can grow more of our own food—in backyards, communal areas, and other open areas. We can plant an array of different kinds of plants, and not just edible ones. One of my favorite examples is living fences. Maya plant trees in a line as a fence. When they are young, they are attached with barbed wire. Once trees mature, however, the barbed wire is no longer necessary. Living fences are a sight to behold. Q'eqchi' Maya villagers in Alta Verapaz, Guatemala, for instance, plant hibiscus (*Hibiscus rosa-sinensis* L.) as living fences, which demarcates boundaries and provides flowers that also have medicinal properties.[23]

To get rid of "pests," instead of using chemicals, we can use, for example, ducks to control them or other natural means. In California, stingless wasps are used to control the destructive (to crops, anyway) apple moth, and owls are used to protect vineyards from rodents. There is so much potential for various entities to connect and relate in different ways. We must take care how we do this—too many times throughout the world introduced species have become invasive species. For example, the spread of killer or "Africanized" bees outside of Africa is an unintended consequence of trying to improve honey production in Brazil beginning in 1956. African queen bees soon escaped, and these aggressive bees have been spreading throughout South America and beyond ever since.

Domesticating animals undoubtedly has benefitted humankind but has led to uncontrolled breeding, methane emissions, and deforestation. There are other costs for changing up the natural order of things. For example, an unintended consequence of our reliance on livestock over time is the increasing number of influenza pandemics due to viruses coevolving with the domestication of pigs and birds.[24] Zoonosis, the transmission of diseases between animals and humans, also accounts for AIDS, Ebola, and probably the coronavirus (SARS-CoV-2).[25]

We can reduce our meat consumption and prevent further transformations of forested areas to grasslands for grazing livestock. This would also decrease the amount of methane, a potent greenhouse gas. The mostly plant-based diet of the Maya sustained them and the environment for millennia. To give you a sense of the magnitude of the impact of livestock, it is estimated that they dominate the total mammal biomass on the Earth, coming in at 694,444,444 tons, followed by humans at 429,894,180 tons, and lastly by terrestrial wild mammals at 22,045,855 tons.[26]

We also need to reevaluate pets, something else humans created through domestication, originally as hunting dogs and for meat consumption, milk, fur, and hides. Domesticated pets only became popular for Western elites in the seventeenth century, and later in the eighteenth century for the middle classes. Today there are millions of pets worldwide—an estimated 900 million dogs, over 400 million cats, and millions of fish and birds.[27] But at what cost in terms of food, water, organic waste, toys, leashes, beds, and impacts on native species? For instance, domestic cats have contributed to the extinction of birds, reptiles, and mammals.[28] Further, 75–85% of dogs are free range, and in the United States alone there are about 75 million feral cats.

I know this solution is a difficult one. They don't choose us; we choose them. We forcibly remove them from their mothers and siblings. We enslave them. Don't get me wrong. I grew up around cats and dogs. I adore pets. But I am trying to present solutions for how to get us and our planet back to a sustainable track. We need to take care of existing pets and stop breeding them.

And then there are zoos that "protect" and breed wild animals—and provide an excuse to take over their native homes. The assumption that animals would be extinct if not for zoos is just a platitude. Instead of zoos, we can protect and promote wildlife in their original homes and/or national parks. We could still see them via cameras—it is so much better to see them interact in their element rather than in cages. Further, evidence indicates that restoring and protecting wild animals and letting them do what they do can boost carbon capture and storage.[29]

And there is no such thing as "wild pets." We created pets. We forcibly capture and cage wild animals.

Green Cities, Urban-Rural Interaction, and Local Networks

Precolonial tropical cities, with their extensive reach across the nonurban landscape, flourished—much more so than premodern Chinese cities, Rome, and Constantinople (Istanbul).[30]

Maya cities had long histories, in some cases over 2,000 years. They included reservoirs and open spaces for urban agricultural fields, orchards, fishponds, gardens, patches of forest with flora and fauna, residences, ball courts, palaces, temples, processional roadways, markets, and plazas that could hold over 10,000 people. Their long histories indicate that the ancestral Maya efficiently disposed of waste, including human waste that they probably used as fertilizer for fields, orchards, and gardens.

Urban-rural interaction was critical because of the scattered distribution of cities, dispersed rural farmsteads, and forested landscapes. Urban and rural systems were linked and formed an enduring partnership. Farmers ultimately survived without cities, but cities could not and did not survive without farmers, since tribute labor, goods, and services funded royal life and infrastructure. The lack of beasts of burden and seasonal river travel meant that ancestral Maya bearers transported items and food mostly from the immediate vicinity. For example, all construction materials used to build thatch residences and monumental temples and palaces were made from locally sourced materials. In brief, they relied on a local resource network and bartering, exchange, and reciprocity.

Urban planning was sustainable—it was the kings who failed. But Maya kings were still around for over 1,000 years, an amazing feat. Ironically, the more kings depended on reservoirs to attract subjects, the more vulnerable they became to decreasing rainfall;

they should have diversified. Political systems come and go, but families persevere. Today we can replace "kings" with "governments" and "transnational corporations." The same thing is happening in the sense of their path-dependent ways with their complete dependency on fossil fuels, for example.

Solutions

We can transform our cities into *yax* ones—green and blue—and integrate them better with the vibrant, pulsating landscape. Maya green cities provide archetypes for urban planning with their monumental buildings interspersed with open and green spaces for gardens, orchards with a variety of tree species and fields, CWs, and fishponds.

The ancestral Maya did not build in flood zones, and neither should we.

We can focus on strengthening urban-rural interaction. We—individuals, communities, cities, states, nations, and transnational corporations—can rely more on local networks for services, products and information to reduce transportation costs, fuel use, pollution, emissions, and the introduction of nonnative species. We can walk or ride bikes instead of driving everywhere. The lack of a major transportation system did not impact the ancestral Maya because they relied on local resources for most everything.

We can rethink our reliance on the global market economy, where value is unpredictable and arbitrary. We can bring back nonmonetary and local bartering, exchange, and reciprocity, where value is determined by interested parties rather than global trends.

Technology and Labor

Without beasts of burden or metals, the ancestral Maya relied on labor and stone tools that left a more sustainable footprint than we see at present, even when the amount of labor used grew in tandem with cities and infrastructure during the Late Classic (c. 600–800 CE). They also relied on technologies that basically did not change for millennia because they worked—agricultural implements (e.g., digging sticks and chert hoes) and agricultural features (e.g., dams and terraces), ceramic manufacture, and stone tool production. And their technologies did not come at the expense of water, soils, forests, and other entities. The Maya kept their extraction of nonhuman entities to only what they needed (chert, clay, temper, game, and flora), thus leaving a sustainable footprint.

We rely too heavily on technology to save the day; our thinking being that it has saved us throughout our history. Technology (e.g., increased automation, "renewable" energy) alone is inadequate over the long run since it runs on *finite* resources and energy. Further, the mining strategies used to extract nonrenewable resources to manufacture car batteries and solar panels puts land, resources, water supplies, and human and nonhuman public health at risk. And most current technologies have short-term benefits and contribute to the out-of-control consumption spiral in which we find ourselves.

Technology, green or not, also has unintended consequences. Take wind turbines. They kill numerous migratory bats already endangered by white-nose syndrome. Bats play a vital role in the chain of life—pollination, seed dispersal, fertilization, and insect control. An even more painful unintended consequence has to do with the cotton gin. Before Eli Whitney invented the cotton gin in 1794, the number of enslaved people in the American South had been declining. All of this changed with the cotton gin and the subsequent increased demands for cotton—and the labor of enslaved people. There is also aviation, which changed how easily people moved over long distances—as well as spreading diseases, as we witnessed with the spread of COVID-19. It also changed warfare, a trend that continues with the current use of unmanned drones to kill people. There are many more examples.

A potential future unintended consequence would be if another Dust Bowl were to take place and blocked sunlight caused by dust clouds lessened the efficiency of solar panels. The expanded use of solar panels will take up extensive amounts of land, which would fill energy needs, but at what cost to nonhuman habitats, biodiversity, cultural heritage, and agricultural production, to name a few? Driverless cars may ultimately result in greater energy expenditures. People, freed from driving, can conduct other activities while in vehicles that require energy, such as communing with smartphones, watching TV, or playing videos. My point is that technology *alone* won't save us unless it is *truly* green.

Solutions

While innovative technological solutions will help, especially in the short term, we don't always need new solutions when we have lessons from the past. We can develop nonexploitative small to large labor projects using technologies that don't rely on electricity or batteries (e.g., hammers, nails, screwdrivers). We can create new jobs, especially for those who may lose their jobs with energy-consuming companies. Labor is green energy. It is renewable and local, and we have a seemingly endless supply of it. In the long run, labor-intensive projects are more durable than technological ones alone. Finally, we can think outside the box; for example, can we use the photosynthesis process to generate power? If so, this would be true solar power.

Multifunctional Infrastructure, Reuse, and Repurpose

The ancestral Maya built and produced multifunctional constructions and items; for instance, raised causeways for walking and processions were also used as dams and to divert and capture water. They repurposed things. They used *metate* and *mano* fragments as hammerstones and pestles. Broken ceramics were used for making net weights for fishing, spindle whorls for spinning cotton and other fibers, circular discs to use as lids, material to line reservoirs, and as construction fill. Ceramic sherds manufactured with volcanic ash temper are like fine-grained sandpaper and can be used for polishing and smoothing.[31] They also reused cut limestone blocks and other architectural elements in construction projects.

In Chiapas, Mexico, the Tzeltal Maya of Chanal and Aguacatenango use broken *metates* and *manos* (some from archaeological sites) to grind calcite temper.[32] They use broken pots to store calcite and clay, for growing plants, and storing grains. They use small body sherds for paving and fill chinking; necks with rims as pot stands and to protect young sprouts; vessel bases to mix lime and as animal feeding dishes and flowerpots; handles as kiln furniture, wall hangers, door handles, and curtain holders; and large sherds as nests for chickens, trays for drying seeds, roof slating, containers for mixing paints, candle holders and stands, and for many other uses.

While working in west-central Belize, one of our excavation assistants from Bullet Tree Falls, Julio, used an empty green wine bottle to catch fish. He had removed the bottom of the bottle, placed a lure in through the nozzle, and put it in the water. Fish go inside but can't get out since they can't swim backward. The result: fresh fish. Electrical fan frames are used as containers to grow orchids and other flowers and plants, *metates* as bird baths or dog food and water containers, and broken ceramic vessels to grow plants. The Choc family uses a hollowed-out tree stump to churn butter (see Figure 1.1). There are so many more examples.

Solutions

We can manufacture and build things that have long-lasting and multipurpose uses. Currently, everything is single purposed. We are a sole-minded society. But we can change how we manufacture and consume. For example, roads can also serve as flood barriers. Raised roads can provide shade and areas for walkways, bike paths, and small businesses. And if some ancient Roman roads are still in use 2,000 years later, surely we can find more efficient ways to construct roads that don't require maintenance every year.[33] Ancient Maya buildings, plazas, and causeways have withstood the test of time due to the natural additives, including plant extracts, Maya used to make limestone plaster and mortar that reduced cracking and calcite dissolution.[34] We can do the same or something similar.

We can reuse and repurpose. Recycling is not enough—recycling consumes lots of water through, for example, melting and reshaping plastics.[35] So don't use plastic! Or at least use biodegradable plastic. The Biodegradable Research Institute (https://bpiworld.org/) provides tested products, and not just plastic ones. Recycling is expensive and requires lots of resources and capital—we can't rely on it alone.

We can turn energy-sucking factories into repair and repurpose centers that would create jobs since they would rely on labor and nonelectric tools.

For parks, more and more urban planners and landscape architects incorporate a multi-purpose approach. Parks can provide several benefits beyond recreation, such as clean water, preventing flooding, serving as wildlife habitats, and growing flowers and food.[36]

Repurposing Organic Waste and the Dead

A Wake
Great Aunt:
There's no end to sickness,
death won't go away.
At least you're not the only one;
I'm going to die too.
All of us will become Earth.
All of us will be mud.
There are no two ways about it:
I'm coming right behind you, here beside you.
Little marigold:
Flower of death:
How many are buried beneath this cross?
How many underneath our prayers?

—MARUCH MÉNDES PÉRES[37]

The ancestral Maya gave back to the tropical forest, for example, through the disposal of human waste and corporeal remains. Remember, souls are recycled, not corporeal remains. Giving back serves several purposes: disposing of departed loved ones, removing human waste, and nourishing plants, trees, insects, and other life.

Archaeologists have not found any obvious latrines at Maya sites or cities. The Maya likely used human waste collected in chamber pots (ceramic jars) as fertilizer in their home gardens, orchards, and *milpas*. Clearly, they handled waste management well, which was critical in cities with tens of thousands of city dwellers and a porous limestone bedrock where waste seepage into reservoirs would have been catastrophic.

In the late 1990s a Peace Corps volunteer obtained an $80,000 grant from an international fund to build outhouses in a Maya village in Belize because she noticed that not many existed. If only she had asked people *why* they did not have outhouses—perhaps night soil, or human excrement typically collected at night, was used as fertilizer? As more villagers acquire wealth, they more and more adopt modern practices and technologies—including indoor plumbing and flushing toilets.

The ancestral Maya did not have cemeteries with rows upon rows of headstones. And since they interred only about 10% of deceased family members in homes and tombs, this means that the Maya placed the remains of about 90% of the dead somewhere else, likely in the jungle, to give back, so to speak (see Chapter 7). Even though archaeologists find ancestral remains in caves (see Chapter 4), they only account for a fraction of the 90% of the deceased not buried under the floors of houses or in shrines, temples, and palaces.

Ethnographer Allen Christenson tells us that it is common for Maya to plant trees over burials. For example, in highland Guatemala, Santiago Atitlán's cemetery looks like an orchard of rows upon rows of fruit trees including *pataxte* (*Theobroma bicolor*), a type of cacao.[38] The Tz'utujil Maya plant trees in mounds of the rich Earth over recently deceased family members. Trees embody their reborn or recycled souls. A similar practice occurred in the early eighteenth century, but in maize fields instead of cemeteries. The corporeal remains nourish trees and maize, and they in turn nourish reborn souls. We also see evidence of this practice in the archaeological record. For example, the decorated sarcophagus of Palenque's most powerful king, K'inich Janaab' Pakal (603–683 CE) in the Temple of Inscriptions depicts not only his ancestors emerging from the Earth, but sprouting fruit trees behind each of the 10 ancestors.

And in Mensabak, Chiapas, Mexico, the Lacandon Maya consider trees in abandoned house lots as belonging to deceased or absent owners.[39] I mentioned earlier how the presumed primary forests we see in some parts of the Maya area today actually represent descendant forests reflecting ancient Maya collaboration. Forests also likely are massive "cemeteries"—a forest of ancestors living among nonhuman entities for eternity.

Solutions

We can recycle organic waste as fertilizer for local fields, forests, parks, gardens, and orchards. Doing so also would remove hazardous materials from settled areas and would save on transportation costs and pollution. Constructed wetlands can provide fertilizer in the form of fish feces and other bottom debris since they require dredging every few years. It is also necessary to harvest and replenish aquatic plants saturated with nitrogen, phosphorus, and other nutrients, which also can be used as fertilizer for fields and gardens.

We can handle household or community-level waste management, for example, by using night soil for local gardens, as traditional farmers have done for ages (e.g., precommunist China, eighteenth century Japan). We can turn energy-sucking factories into natural sewage treatment centers—and create new jobs that turn sewage into fertilizers (biosolids). To get a better sense of how we can repurpose human waste, take a look at "Ten Ways that Human Waste Isn't Waste at All" by Nicholas C. Kawa.[40] It also can be used to generate energy through capturing the methane gases human waste creates.

We can place our loved ones in natural burials—no monuments, no markers, no plots, no embalming, no vaults, and no metal caskets and fittings. "Become soil when you die."[41] Caskets are not required by law in the United States. We instead can bury loved ones and plant trees like the Maya did. Funeral homes can change their business strategy and use, for example, burial pods. We can make cremations more green,[42] for example, by using aquamation, "a gentle process that uses water instead of fire to return a body back to mother nature"—it uses 90% less energy than cremations and has no direct greenhouse gas emissions.[43] We can perform burials at sea, create memorial reefs, conduct home funerals where the body is prepared at home, and if people feel caskets are necessary, use local and plain wood ones for backyard burials, in rural areas at least. We can use banana leaves or bamboo and other more sustainable materials, or even living coffins that can grow and nourish mushrooms.[44] The death of loved ones can contribute to the cycle of life, nourishing nonhuman entities that play essential roles in world maintenance. For instance, Return Home is a company in Auburn, Washington, that "gently transforms human remains into nutrient-rich soil that can be used to promote new life."[45]

We can plant trees over graves and create family forests. Bodies create trees of life. Death begets life. Bodies nourish trees, soil, insects, and many other entities. We can plant millions of trees this way and reforest areas. Lineage forests can memorialize loved ones while promoting forestation and its concomitant benefits. After all, who would cut down their family trees?

Migration

Migration has always been part of human history since the dawn of humanity. However, the impacts of migration changed after the divorce of nature and culture. Instead of immigrants learning from

Indigenous peoples and nonhumans already living in an area, they imposed their ideas and practices from their former homeland. In many cases, the immigrant (student) became the colonizer (imposer), with devastating repercussions still felt today. In the United States and other countries with colonial histories, our unsustainable path began when immigrants/colonizers brought with them from Europe techniques and ways of viewing the world—and did not learn how to engage with the environment from the Indigenous inhabitants.

Maya migration. It happened in the past and is happening now. As families and communities grow, one option to avoid overusing water, soils, and forests is to split off and start a new farmstead and community elsewhere. Some farmers intensified agricultural practices via terraces, dams, and ditches, like many did in the Late Classic (c. 600–800 CE). They emigrated in masse from the interior southern lowlands between 800 and 900 CE during the multiple prolonged droughts and found new homes and economic opportunities (e.g., maritime trade) elsewhere.

The ancestral Maya did what they had to do for the survival of their families and their environment. They adapted. When necessary, they left their homes and ancestors to ensure the survival of humans and nonhumans. Today the Maya migrate to escape violence, poverty, and the impacts of a changing climate.

Solutions

Families "leave to live."[46] Migration is more challenging than in the past due to strict borders, prejudices, political disputes, and economic fears. And while the first immigrants to the Americas learned from established inhabitants, including animals, birds, trees, water, and other nonhumans, immigrants today can contribute to their new homes through their skills, knowledge (e.g., knowledge of medicinal plants and professional expertise), experiences, drive, and motivations—and ideas about how to diversify strategies in their new home. Oftentimes xenophobic anti-immigrant people forget their own immigrant histories.

Migration needs to continue, and with greater urgency. Climate refugees are a fact of life today and will be for the foreseeable future and must be part of any type of future planning, an issue all countries must face, address, accept, and assist. Nations all over the globe need to rethink migration moving forward because the number of climate refugees will grow as the impacts of global climate change become more drastic.

Final Remarks

There are many other things we can do for the survival of our "family" and planet. The solutions presented here are just a beginning. For instance, to ease our human imprint, we can practice voluntary family planning and ensure access to education, contraceptives, and abortion for all.

These solutions will not make anyone wealthier or increase profit margins per se. But they will help us survive in an increasingly changing world. In doing so, what will increase is the satisfaction that you are making a difference. More people in the United States, for example, need to vote and hold politicians to their promises—especially since most do not fulfill them. People, communities, and groups can apply pressure. Presently, "family values" only cover family members parents can interact with in the present rather than future descendants they cannot interact with. And we also need the best of both top-down (governments and transnational corporations) and bottom-up (nearly everything else) action to make a difference. Every one of us can and must act.

Current hot topics like abortion and gun control that focus on anthropocentric individual rights need to be replaced with the same passion and energy for our future—reforesting, rewilding, addressing climate change, and devising sustainable strategies.

Current plans to address climate change in sustainable ways privilege humans. They are anthropocentric, short-term, and rely too much on technology. To take care of our entire family, we need to take care of the environment and our planet. The family or household provides the basic unit of society. It also serves as the basic unit of survival, action, and change. And a key thing to remember is that everything is connected on this planet.

> Let's reconnect.
> Let's move beyond self-interest.
> Let's shake up the hierarchy and not privilege humans.
> Let's embrace an inclusive worldview.
> Let's transform "us versus them" to "us and them" to "us."
> Let's redefine our relationship with nonhumans.
> Let's redefine resources as entities instead of commodities.
> Let's lessen detrimental interaction with resources/entities.
> Let's wean ourselves from materialism and this consumption craze we're on.
> Let's develop nontechnological innovations in addition to technological ones.
> Let's take responsibility for saving our planet.

This book is a call to arms. Step one: hope. Step two: action from all of us. Step three: repeat.

***

In the 500 years since the conquest of the Maya and all that came with it, one of the most detrimental results has been the weakening or loss of an inclusive worldview due to forced conversion to Christianity and continued missionization. Maya sustainable practices are being replaced by modern ones—to the detriment of the environment and sustainability. In the face of their current place in society and the need to take care of their families, they don't have much choice, especially due to the large-scale selling of forests to foreign and non-Maya concerns. Most non-Maya practices have short-term benefits—especially massive deforestation and monocropping where Indigenous input is nonexistent. By ignoring traditional knowledge, the powers that be are decreasing their profit margins and destroying the environment—at the peril of us all.

We are at the edge of a precipice. It is up to all of us whether we go off the cliff like lemmings or turn back and forge a new path. I hope the Maya have shown you the way by opening your eyes to a different way of seeing and interacting with the world to ensure a sustainable future.

ENDNOTES

Preface

1. https://scholastic.nd.edu/issues/no-planet-b-ban-ki-moon-discusses-global-citizenship/

2. R. Cooke, F. Sayol, T. Andermann, T. M. Blackburn, M. J. Steinbauer, A. Antonelli, and S. Faurby. 2023. Undiscovered Bird Extinctions Obscure the True Magnitude of Human-Driven Extinction Waves. *Nature Communication* 14:8116. https://doi.org/10.1038/s41467-023-43445-2; R. van Klink, D. E. Bowler, K. B. Gongalsky, M. Shen, S. R. Swengel, and J. M. Chase. 2023. Disproportionate Declines of Formerly Abundant Species Underlie Insect Loss. *Nature* 628:359–364. https://doi.org/10.1038/s41586-023-06861-4.

3. S. Martin and N. Grube. 2008. *Chronicle of the Maya Kings and Queens.* 2nd ed. Thames and Hudson, New York, p. 30.

Introduction

1. I have permission to use his real name, as well as the names of other foremen and excavation assistants. Cleofo has worked with me since 1998.

2. K. Emery. 2007. Assessing the Impact of Ancient Maya Animal Use. *Journal for Nature Conservation* 15:184–195.

3. G. Ceballos and P. R. Ehrlich. 2023. Mutilation to the Tree of Life via Mass Extinction of Animal Genera. *Proceedings of the National Academy of Sciences* 120:e2306987120. https://doi.org/10.1073/pnas.2306987120.

4. A. Leopold. 1991 [1942]. Land-Use and Democracy. In *The River of the Mother of God: And Other Essays by Aldo Leopold,* edited by S. L. Flader and J. B. Callicott, pp. 295–300. University of Wisconsin, Madison.

5. Organized crime began during the early 20th century labor wars, and their power grew as they increasingly controlled labor unions; J. B. Jacobs. 2019. The Rise and Fall of Organized Crime in the United States. *Crime & Justice* 49:17–67.

6. For a nice and succinct summary of this transformation, see C. Merchant. 2005. *Radical Ecology: The Search for a Livable World.* 3rd ed. Routledge, New York, pp. 41–61.

7. R. W. Kimmerer. 2021. A Family Reunion Near the End of the World. In *Kinship: Belonging in a World of Relations.* Vol. I, *Planet,* edited by G. Van Horn, R. W. Kimmerer, and J. Hausdoerffer, pp. 111–124. Center for Humans and Nature Press, Libertyville, IL, p. 115.

8. V. Watts. 2013. Indigenous Place-Thought and Agency amongst Humans and Nonhumans (First Woman and Sky Woman go on a European World Tour!). *Decolonization: Indigeneity, Education and Society* 2:20–34, p. 25. In the Haudenosaunee (Iroquois) creation history Sky Woman fell through a hole in the sky and was helped to earth by different kinds of birds and landed on the back of a turtle. Together with animals, she created lands and humankind. For more details, see https://www.oneidaindiannation.com/the-haudenosaunee-creation-story/.

9. Watts, p. 24.

10. Kimmerer, Family Reunion, p. 112.

11. See P. Roberts. 2019. *Tropical Forest Prehistory, History, and Modernity.* Oxford University Press, London, pp. 1–24.

12. P. W. Staten, J. Lu, K. M. Grise, S. M. Davis, and T. Birner. 2018. Re-examining Tropical Expansion. *Nature Climate Change* 8:768–775. https://doi.org/10.1038/s41558-018-0246-2.

13. Maya can see the remains of their royal forebears today in books and museums. Before the Belize Museum opened in 2002 in Belize City, much of the Belize archaeology collection was stored in the basement of the original Department of Archaeology building in Belmopan. The collection included confiscated looted items and legally excavated artifacts.

14. A. Ford and R. Nigh. 2015. *The Maya Forest Garden: Eight Millennia of Sustainable Cultivation of the Tropical Woodlands.* Left Coast Press, Walnut Creek, CA. See also T. S. Beach, S. Luzzadder-Beach, N. Dunning, J. Jones, J. Lohse, T. Guderjan, S. Bozarth, S. Millspaugh, and T. Bhattacharya. 2009. A Review of Human and Natural Changes in Maya Lowland Wetlands over the Holocene. *Quaternary Science Reviews* 28:1710–1724; T. S. Beach,

S. Luzzadder-Beach, D. Cook, N. Dunning, D. J. Kennett, S. Krause, R. Terry, D. Trein, and F. Valdez. 2015. Ancient Maya Impacts on the Earth's Surface: An Early Anthropocene Analog? *Quaternary Science Reviews* 124:1–30; N. P. Dunning, T. Beach, and S. Luzzadder-Beach. 2012. Kax and Kol: Collapse and Resilience in Lowland Maya Civilization. *Proceedings of the National Academy of Sciences* 106:3652–3657; C. L. McNeil, D. A. Burney, and L. P. Burney. 2010. Evidence Disputing Deforestation as the Cause for the Collapse of the Ancient Maya Polity of Copan, Honduras. *Proceedings of the National Academy of Sciences* 107:1017–1022.

15. Visit the VOPA project webpage for field reports, photos, videos, and publications: https://publish.illinois.edu/valleyofpeace/.

16. Prior to the establishment of the Valley of Peace Village, archaeologists from the Belize Department of Archaeology (now the Institute of Archaeology) conducted an archaeological survey and excavations to assess the impact of constructing the village on Maya sites.

Chapter 1

1. W. F. Hanks. 1990. *Referential Practice: Language and Lived Space among the Maya*. University of Chicago Press, Chicago, p. 389.

2. R. W. Kimmerer. 2013. *Braiding Sweetgrass: Indigenous Wisdom, Scientific Knowledge, and the Teachings of Plants*. Milkweed Editions, Minneapolis, MN, p. 169.

3. K. Brandon. 2014. *Ecosystem Services from Tropical Forests: Review of Current Science*. Center for Global Development Working Paper 380, Washington, DC, p. 17.

4. E. A. Orijemie. 2018. Conserving the Tropical Rainforests in Nigera: In Whose Interest? In *Exploring Frameworks for Tropical Forest Conservation: Integrating Natural and Cultural Diversity for Sustainability: A Global Perspective*, edited by N. Sanz, D. Rommens, and J. Pulido Mata, pp. 216–227. UNESCO Mexico, Mexico City, p. 220.

5. X. Cai, W. J. Riley, Q. Zhu, J. Tang, Z. Zeng, G. Bisht, and J. T. Randerson. 2019. Improving Representation of Deforestation Effects on Evapotranspiration in the E3SM Land Model. *Journal of Advances in Modeling Earth Systems* 11:2412–2427. https://doi.org/10.1029/2018MS001551.

6. C. Smith, J. C. A. Baker, and D. V. Spracklen. 2023. Tropical Deforestation Causes Large Reductions in Observed Precipitation. *Nature* 615:270–275. https://doi.org/10.1038/s41586-022-05690-1.

7. For example, there are bananas (apple bananas and regular bananas), plantains, lime, cacao, coconut, orange, maley apple, crabbo, guava, soursop, custard, bird pepper, cilantro, oregano, avocado, mango, breadfruit (ramón nut), and black orchids.

8. C. Lindsay. 2011. *Culturally Modified Landscapes from Past to Present: Yalbac, Belize*. M.A. paper. University of Illinois at Urbana-Champaign. http://publish.illinois.edu/valleyofpeace/files/2019/07/Lindsay20MA20 official20copy.pdf; see also N. J. Ross. 2011. Modern Tree Species Composition Reflects Ancient Maya "Forest Gardens" in Northwest Belize. *Ecological Applications* 21:75–84; A. Ford, G. Turner, and H. Mai. 2023. Favored Trees of the Maya Milpa Forest Garden Cycle. In *Ecotheology: Sustainability and Religions of the World*. IntechOpen. https://doi.org/10.5772/intechopen.106271.

9. The Mayan name of the Maize God remains elusive, though there are references or aspects, or even different Maize Gods, mentioned in the inscriptions (Juun Ixi'm or One Maize/Tonsured Maize God and Waxak Ajan or Foliated Maize God). See K. Taube. 1992. *Major Gods of Ancient Yucatan*. Studies in Pre-Columbian Art and Archaeology no. 32. Dumbarton Oaks, Washington, DC, pp. 44–50; S. Martin. 2012. Hieroglyphs from the Painted Pyramid: The Epigraphy of Chiik Nahb Structure Sub 1–4, Calakmul, Mexico. In *Maya Archaeology 2*, edited by C. Golden, S. D. Houston, and J. Skidmore, pp. 60–81. Precolumbia Mesoweb Press, San Francisco; D. Stuart. 2006. The Language of Chocolate: References to Cacao on Classic Maya Drinking Vessels. In *Chocolate in Mesoamerica: A Cultural History of Cacao*, edited by C. McNeil, pp. 184–201. University Press of Florida, Gainesville.

10. C. L. Hogue. 1993. *Latin American Insects and Entomology*. University of California Press, Berkeley, p. 180.

11. D. L. Martin and A. H. Goodman. 2002. Health Conditions before Columbus: Paleopathology of Native North Americans. *Western Journal of Medicine* 176:64–68; A. F. Ramenofsky, A. K. Wilbur, and A. C. Stone. 2003. Native American Disease History: Past, Present and Future Directions. *World Archaeology* 35:241–257.

12. T. Garel and S. Matola. 1996. *A Field Guide to the Snakes of Belize*. Belize Zoo and Tropical Education Center, Belize.

13. https://www.epa.gov/enviroatlas/enviroatlas-benefit-category-biodiversity-conservation

14. See C. A. Kray. 2023. *Maya-British Conflict at the Edge of the Yucatecan Caste War*. University Press of Colorado, Louisville.

15. Due to the COVID shutdown, field plans were cancelled in 2020 and 2021, which meant at least 100 years of Maya family histories were lost to plowing.

16. L. J. Lucero, R. Taylor, Y. Wang, and L. J. Kosakowsky. 2023. Denuded Landscapes and Exposed Neighborhoods: Results of the 2022 Valley of Peace Archaeology Project. *Research Reports in Belizean Archaeology* 18:289–299.

Chapter 2

1. A. J. Christenson. 2007. *Popol Vuh: The Sacred Book of the Maya.* University of Oklahoma Press, Norman, p. 71.

2. J. T. Larmon, H. G. McDonald, S. Ambrose, L. R. G. Desantis, and L. J. Lucero. 2019. A Year in the Life of a Giant Ground Sloth During the Last Glacial Maximum in Belize. *Science Advances* 5:eaau1200.

3. Some scholars propose that a mix of hunting and the end of the last ice age (warmer and drier) together led to the extinction of megafauna in the Americas by c. 10,000 years ago; see D. J. Meltzer. 2020. Overkill, Glacial History, and the Extinction of North America's Ice Age Megafauna. *Proceedings of the National Academy of Sciences* 117:28555–28563. https://www.pnas.org/doi/full/10.1073/pnas.2015032117. Scholars in another study add human-caused fires as altering the landscape enough to impact megafauna; see F. R. O'Keefe, R. E. Dunn, E. M. Weitzel, M. R. Waters, L. N. Martinez, W. J. Binder, J. R. Southon, et al. 2023. Pre-Younger Dryas Megafaunal Extirpation at Rancho La Brea Linked to Fire-Driven State Shift. *Science* 381:6659. https://doi.org/10.1126/science.abo3594.

4. R. W. Kimmerer. 2013. *Braiding Sweetgrass: Indigenous Wisdom, Scientific Knowledge, and the Teachings of Plants.* Milkweed Editions, Minneapolis, MN, p. 205. Anishinaabe include the Ojibwe, Ottawa, and Potawatomi Native American tribal nations.

5. See, for example, R. Cobos, G. de Anda Alanís, and R. García Moll. 2014. Ancient Climate and Archaeology: Uxmal, Chichén Itzá, and Their Collapse at the End of the Terminal Classic Period. In *The Resilience and Vulnerability of Ancient Landscapes: Transforming Maya Archaeology through IHOPE*, edited by A. Chase and V. Scarborough, pp. 56–71. Archeological Papers of the American Anthropological Association no. 24. Wiley-Blackwell, Hoboken, NJ; C. Isendahl, N. P. Dunning, and J. A. Sabloff. 2014. Growth and Decline in Classic Maya Puuc Political Economies. In *The Resilience and Vulnerability of Ancient Landscapes: Transforming Maya Archaeology through IHOPE*, edited by A. Chase and V. Scarborough, pp. 43–55. Archeological Papers of the American Anthropological Association no. 24. Wiley-Blackwell, Hoboken, NJ.

6. For the history of the adoption and spread of domesticated plants and traditional Maya agricultural practices, see C. Cagnato. 2021. Gathering and Sowing across the Central Maya Lowlands: A Review of Plant Use by Preceramic Peoples and the Early to Middle Preclassic Maya. *Ancient Mesoamerica* 32:486–501; S. Morell-Hart, L. Dussol, and S. L. Fedick. 2022. Agriculture in the Ancient Maya Lowlands (Part 1): Paleoethnobotanical Residues and New Perspectives on Plant Management. *Journal of Archaeological Research* 31:561–615. https://doi.org/10.1007/s10814-022-09180-w; and S. L. Fedick, S. Morell-Hart, and L. Dussol. 2023. Agriculture in the Ancient Maya Lowlands (Part 2): Landesque Capital and Long-term Resource Management Strategies. *Journal of Archaeological Research* 32:103–154. https://doi.org/10.1007/s10814-023-09185-z.

7. D. Stuart, H. Hurst, B. Beltrán, and W. Saturno. 2022. An Early Maya Calendar Record from San Bartolo, Guatemala. *Science Advances* 8:eabl9290.

8. S. L. Fedick. 1996. An Interpretive Kaleidoscope: Alternative Perspectives on Ancient Agricultural Landscapes of the Maya Lowlands. In *The Managed Mosaic: Ancient Maya Agriculture and Resource Use*, edited by S. L. Fedick, pp. 107–131. University of Utah Press, Salt Lake City; B. L. Turner II and D. Lawrence. 2012. Land Architecture in the Maya Lowlands: Implications for Sustainability. In *Biodiversity in Agriculture: Domestication, Evolution, and Sustainability*, edited by P. Gepts, T. R. Famula, R. L. Bettinger, S. B. Brush, A. B. Damania, P. E. McGuire, and C. O. Qualset, pp. 445–463. Cambridge University Press, Cambridge, UK.

9. See, for example, Part VII in S. Hutson and T. Ardren (editors). 2020. *The Maya World*. Routledge, London; M. A. Masson and D. A. Freidel. 2012. An Argument for Classic Era Maya Market Exchange. *Journal of Anthropological Archaeology* 31:455–484; and J. A. Sabloff. 2007. It Depends on How you Look at Things: New Perspectives on the Postclassic Period in the Northern Maya Lowlands. *Proceedings of the American Philosophical Society* 151:11–25.

10. R. J. Sharer and L. Traxler. 2006. *The Ancient Maya*. 6th ed. Stanford University Press, Stanford, CA, pp. 757–778.

11. See L. J. Lucero. 2006. *Water and Ritual: The Rise and Fall of Classic Maya Rulers*. University of Texas Press, Austin.

12. A. M. Tozzer. 1941. *Landa's Relación de Los Cosas de Yucatán*. Papers of the Peabody Museum of American Archaeology and Ethnology, no. 28. Harvard University, Cambridge, MA, p. 42.

13. Kaqchikel and K'iche' are different languages and ethnic groups.

14. P. McAnany. 2020. Imagining a Maya Archaeology That Is Anthropological and Attuned to Indigenous Cultural Heritage. *Heritage* 3:318–330.

15. See V. Sanford. 2003. *Buried Secrets: Truth and Human Rights in Guatemala.* Palgrave Macmillan, London.

Chapter 3

1. A. J. Christenson. 2007. *Popol Vuh: The Sacred Book of the Maya.* University of Oklahoma Press, Norman, p. 73.

2. S. Harding and A. Penny (editors). 2020. *State of the Tropics 2020 Report.* James Cook University, Douglas, Australia, p. 13.

3. H. Yang, G. Lohmann, J. Lu, E. J. Gowan, X. Shi, J. Liu, and Q. Wang. 2020. Tropical Expansion Driven by Poleward Advancing Midlatitude Meridional Temperature Gradients. *Journal of Geophysical Research: Atmospheres* 125:e2020JD033158. https://doi.org/10.1029/2020JD033158.

4. United Nations. 2023. *The United Nations World Water Development Report 2023: Partnerships and Cooperation for Water.* UNESCO, Paris, p. 203. https://www.unesco.org/reports/wwdr/2023/en.

5. I have permission from her mother, Nala Teck Choc, to use this quote (Facebook Messenger, February 18, 2023).

6. With assistance from Cleofo, Ernesto, diver and photographer Tony Rath, cave exploration diver Chip Petersen, and underwater videographer Marty O'Farrell.

7. N. P. Dunning, J. Brewer, C. Carr, A. A. Hernández, T. Beach, J. Chmilar, L. G. Sierra, et al. 2022. Harvesting *Ha*: Ancient Water Collection and Storage in the Elevated Interior Region of the Maya Lowlands. In *Sustainability and Water Management in the Maya World and Beyond*, edited by J. T. Larmon, L. J. Lucero, and F. Valdez, Jr., pp. 13–51. University Press of Colorado, Louisville.

8. B. B. Faust, A. Anaya-Hernández, and H. Geovannini-Acuña. 2012. Reclaiming the Past to Respond to Climate Change: Mayan Farmers and Ancient Agricultural Techniques in the Yucatan Peninsula of Mexico. In *Climate Change and Threatened Communities: Vulnerability, Capacity, and Action*, edited by A. P. Castro, D. Taylor, and D. W. Brokensha, pp. 139–151. Practical Action Publishing, Rugby, UK.

9. T. C. Winegard. 2019. *The Mosquito: A Human History of Our Deadliest Predator.* Dutton, New York, pp. 145–146.

10. D. L. Martin and A. H. Goodman. 2002. Health Conditions Before Columbus: Paleopathology of Native North Americans. *Western Journal of Medicine* 176:64–68; A. F. Ramenofsky. 2003. Native American Disease History: Past, Present and Future Directions. *World Archaeology* 35:241–257.

11. Information about Maya reservoirs and constructed wetlands is summarized from L. J. Lucero. 2023. Ancient Maya Reservoirs, Constructed Wetlands, and Future Water Needs. *Proceedings of the National Academy of Sciences* 120:e2306870120. https://doi.org/10.1073/pnas.2306870120.

12. U.S. Environmental Protection Agency, https://www.epa.gov/wetlands/constructed-wetlands.

13. I mostly rely on the FLAAR Mesoamerica website (https://flaarmesoamerica.org/) for information about wetland biota.

14. The library has been renamed Water Resources Collections and Archives and has been relocated to UC Riverside and California State University San Bernardino: https://library.ucr.edu/collections/water-resources-collections-archives.

15. R. D. Hansen, S. Bozarth, J. Jacob, D. Wahl, and T. Schreiner. 2002. Climatic and Environmental Variability in the Rise of Maya Civilization: A Preliminary Perspective from Northern Peten. *Ancient Mesoamerica* 13:273–295; R. D. Hansen, C. Morales-Aguilar, J. Thompson, R. Ensley, E. Hernández, T. Schreiner, E. Suyuc-Ley, and G. Martínez. 2020. LiDAR Analyses in the Contiguous Mirador-Calakmul Karst Basin, Guatemala: An Introduction to New Perspectives on Regional Early Maya Socioeconomic and Political Organization. *Ancient Mesoamerica* 1–40. https://doi.org/10.1017/S0956536122000244.

16. M. Medina–Elizalde, S. J. Burns, J. M. Polanco-Martínez, T. Beach, F. Lases-Hernández, C.-C. Shen, and H.-C. Wang. 2016. High-Resolution Speleothem Record of Precipitation from the Yucatan Peninsula Spanning the Maya Preclassic Period. *Global and Planetary Change* 138:93–102; D. L. Lentz, T. L. Hamilton, N. P. Dunning, V. L. Scarborough, T. P. Luxton, A. Vonderheide, E. J. Tepe, et al. 2020. Molecular Genetic and Geochemical Assays Reveal Severe Contamination of Drinking Water Reservoirs at the Ancient Maya City of Tikal. *Scientific Reports* 10:10316. https://www.nature.com/articles/s41598-020-67044-z.

17. D. Webster. 2014. Maya Drought and Niche Inheritance. In *The Great Maya Droughts in Cultural Context: Case Studies in Resilience and Vulnerability*, edited by G. Iannone, pp. 333–358. University Press of Colorado, Boulder.

18. D. L. Lentz, T. L. Hamilton, N. P. Dunning, E. J. Tepe, V. L. Scarborough, S. A. Meyers, L. Grazioso, and A. A. Weiss. 2021. Environmental DNA Reveals Arboreal Cityscapes at the Ancient Maya Center of Tikal. *Nature Scientific Reports* 11:12725. https://doi.org/10.1038/s41598-021-91620-6.

Chapter 4

1. A. J. Christenson. 2007. *Popol Vuh: The Sacred Book of the Maya*. University of Oklahoma Press, Norman, p. 74.

2. I summarize and expand ideas from L. J. Lucero. 2018. A Cosmology of Conservation in the Ancient Maya World. *Journal of Anthropological Research* 74:327–359.

3. R. J. Sharer and L. Traxler. 2006. *The Ancient Maya*. 6th ed. Stanford University Press, Stanford, CA, p. 720.

4. The cargo system began early in the colonial period as a result of Spanish missionization. For Maya who still participate, men take turns holding year-long posts in one of the several hierarchical levels, each with civic and ceremonial obligations; J. K. Chance and W. B. Taylor. 1985. Cofradías and Cargos: An Historical Perspective on the Mesoamerican Civil-Religious Hierarchy. *American Ethnologist* 12:1–26.

5. E. Z. Vogt. 1969. *Zinacantan: A Maya Community in the Highlands of Chiapas*. Belknap Press of Harvard University Press, Cambridge, MA, p. 521.

6. https://weaving-for-justice.org/maya-symbols-in-the-textiles/

7. R. W. Kimmerer. 2013. *Braiding Sweetgrass: Indigenous Wisdom, Scientific Knowledge, and the Teachings of Plants*. Milkweed Editions, Minneapolis, MN, pp. 55–56.

8. A. Leopold. 1991 [1939]. A Biotic View of the Land. In *The River of the Mother of God: And Other Essays by Aldo Leopold*, edited by S. L. Flader and J. B. Callicott, pp. 266–271. University of Wisconsin, Madison, p. 268.

9. L. K. Pharo. 2007. The Concept of "Religion" in Mesoamerican Languages. *Numen* 54:28–70.

10. A. Balsanelli. 2019. In Search of the Mayan Animal Master: A Short Review. *Global Journal of Archaeology and Anthropology* 8(5):555749. https://doi.org/10.19080/GJAA.2019.08.555749.

11. A. J. Christenson. 2016. *The Burden of the Ancients: Maya Ceremonies of World Renewal from the Pre-Columbian Period to the Present*. University of Texas Press, Austin, p. 326.

12. M. D. Coe. 2011. *The Maya*. 8th ed. Thames and Hudson, New York, p. 224.

13. R. L. Roys. 1933. *The Book of Chilam Balam of Chumayel*. Carnegie Institution of Washington, Washington, DC, p. 99.

14. M. Sahlins. 2022. *The New Science of the Enchanted Universe: An Anthropology of Most of Humanity*. With the assistance of F. B. Henry, Jr., Princeton University Press, Princeton, NJ, p. 119.

15. D. Stuart. 2020. Maya Time. In *The Maya World*, edited by S. Hutson and T. Ardren, pp. 624–636. Routledge, London.

16. This poem and prayer and others quoted from this book were written by Tzotzil Maya women in the late 1990s and show how traditions endure. Á. Past with X. Guzmán Bakbolom and X. Ernandes. 2005. *Incantations: Song, Spells and Images by Mayan Women*. Cinco Puntos Press, El Paso, TX, p. 158.

17. V. Turner. 1973. The Center Out There: Pilgrim's Goal. *History of Religions* 12:191–230.

18. https://www.the-sun.com/news/954051/mob-burn-alive-medicine-expert-witchcraft/; https://www.theguardian.com/global-development/2021/oct/25/guatemala-mayan-spirituality-herbalist-murder-domingo-choc-che

19. Much of the discussion of witchcraft persecution and Maya witches is from L. J. Lucero and S. A. Gibbs. 2007. The Creation and Sacrifice of Witches in Classic Maya Society. In *New Perspectives on Human Sacrifice and Ritual Body Treatments in Ancient Maya Society*, edited by V. Tiesler and A. Cucina, pp. 45–73. Springer Press, New York.

20. E. Oster. 2004. Witchcraft, Weather and Economic Growth in Renaissance Europe. *Journal of Economic Perspectives* 18:215–228.

21. J. Nash. 1970. *In the Eyes of Ancestors: Belief and Behavior in a Mayan Community*. Waveland Press, Prospect Heights, IL, p. 244.

22. E. Z. Vogt. 2004. Daily Life in a Highland Community: Zinacantan in Mid-Twentieth Century. In *Ancient Maya Commoners*, edited by J. C. Lohse and F. Valdez, Jr., pp. 23–47. University of Texas Press, Austin, p. 30.

23. Ethnographer Christine Eber, personal communication, 2005. See also C. Eber. 2000. *Women and Alcohol in a Highland Maya Town: Water of Hope, Water of Sorrow*. 2nd ed. University of Texas Press, Austin; and C. Eber co-authored with "Antonia." 2011. *The Journey of a Tzotzil-Maya Woman of Chiapas, Mexico: Pass Well Over the Earth*. University of Texas Press, Austin.

24. R. J. McGee. 1998. The Lacandon Incense Burner Renewal Ceremony: Termination and Dedication Ritual among the Contemporary Maya. In *The

Sowing and the Dawning: Termination, Dedication, and Transformation in the Archaeological and Ethnographic Record of Mesoamerica, edited by S. B. Mock, pp. 41–46. University of New Mexico Press, Albuquerque.

25. Relatively few Classic Maya kings were powerful enough to have power over life and death. Those sacrificed, based on inscriptions, included captured kings or warriors who were considered valuable, or they wouldn't be sacrificed to gods and ancestors; after all, for it to be a sacrifice, it must be valuable. It's not possible to identify in the archaeological record whether those sacrificed by commoners were slaves, a child purchased from an impoverished family, or criminals.

26. See D. Zhang, H. F. Lee, C. Wang, B. Li, Q. Pei, J. Zhang, and Y. An. 2011. The Causality Analysis of Climate Change and Large-Scale Human Crisis. *Proceedings of the National Academy of Sciences* 108:17296–17301; and E. Oster. 2004. Witchcraft, Weather and Economic Growth in Renaissance Europe. *Journal of Economic Perspectives* 18:215–228.

Chapter 5

1. E. N. Anderson with A. Dzib Xihum de Cen, F. Medina Tzuc, and Pastor Valdez Chale. 2005. *Political Ecology in a Yucatec Maya Community*. University of Arizona Press, Tucson, pp. 116, 118–119.

2. D. C. Maya-Cortés, J. de Dios Figueroa Cárdenas, M. G. Garnica-Romo, R. A. Cuevas-Villanueva, R. Cortés-Martínez, J. J. Véles-Medina, and H. Eduardo Martínez-Flores. 2010. Whole-Grain Corn Tortilla Prepared Using an Ecological Nixtamalisation Process and its Impact on the Nutritional Value. *International Journal of Food Science & Technology* 45:23–28.

3. S. L. Fedick. 2010. The Maya Forest: Destroyed or Cultivated by the Ancient Maya? *Proceedings of the National Academy of Sciences* 107:953–954. https://doi.org/10.1073/pnas.0913578107.

4. E. Z. Vogt. 1969. *Zinacantan: A Maya Community in the Highlands of Chiapas*. Belknap Press of Harvard University Press, Cambridge, MA, p. 146.

5. Most of the information about pilgrimage and Cara Blanca pools is summarized from L. J. Lucero and A. Kinkella. 2015. Pilgrimage to the Edge of the Watery Underworld: An Ancient Maya Water Temple at Cara Blanca, Belize. *Cambridge Archaeological Journal* 25:163–185; and L. J. Lucero, J. Harrison, J. Larmon, Z. Nissen, and E. Benson. 2016. Prolonged Droughts, Short-Term Responses and Diaspora: The Power of Water and Pilgrimage at the Sacred Cenotes of Cara Blanca, Belize. *WIREs Water*. https://doi.org/10.1002/wat2.1148.

6. D. T. Price, V. Tiesler, and C. Freiwald. 2019. Place of Origin of the Sacrificial Victims in the Sacred Cenote, Chichén Itzá, Mexico. *American Journal of Physical Anthropology* 170:98–115. https://doi.org/10.1002/ajpa.23879.

7. For related publications (e.g., Kinkella 2009; Larmon 2019; Lucero and Kinkella 2015; Lucero J. Harrison, J. Larmon, Z. Nissen, and E. Benson. 2016) download from VOPA project website: https://publish.illinois.edu/valley ofpeace/.

8. The Yalbac Ranch property was sold to a consortium of conservation groups headed by The Nature Conservancy in December 2020, who then turned it over to the Belize Maya Forest Trust based in Belize and directed by Dr. Elma Kay; see https://www.rainforesttrust.org/our-impact/rainforest-news/partnering-to-preserve-belize-maya-forest/.

9. See the VOPA project website: https://publish.illinois.edu/valleyofpeace/.

10. R. L. Roys. 1933. *The Book of Chilam Balam of Chumayel*. Carnegie Institution of Washington, Washington, DC, p. 175.

11. Tony Rath, https://www.tonyrath.com/index; Chip Petersen, former owner, Belize Diving Services, Caye Caulker.

Chapter 6

1. A. J. Christenson. 2007. *Popol Vuh: The Sacred Book of the Maya*. University of Oklahoma Press, Norman, pp. 194–195.

2. Christenson, pp. 59–60.

3. Christenson, pp. 65–66.

4. R. W. Kimmerer. 2013. *Braiding Sweetgrass: Indigenous Wisdom, Scientific Knowledge, and the Teachings of Plants*. Milkweed Editions, Minneapolis, MN, pp. 132, 134.

5. S. L. Fedick. 2010. The Maya Forest: Destroyed or Cultivated by the Ancient Maya? *Proceedings of the National Academy of Sciences* 107:953–954. https://doi.org/10.1073/pnas.0913578107.

6. M. A. Canuto, F. Estrada-Belli, T. G. Garrison, S. D. Houston, M. J. Acuña, M. Kováč, D. Marken, et al. 2018. Ancient Lowland Maya Complexity as Revealed by Airborne Laser Scanning of Northern Guatemala. *Science* 361:eaau0137.

7. Scott Fedick, personal communication, 2007.

8. R. Taylor. 2024. *The Health and Biodiversity of the Ancestral Maya Forest: A Zooarchaeological Perspective*. Ph.D. dissertation, University of Illinois at Urbana-Champaign.

9. A. R. Wyatt. 2023. "An Instrument of Grace": Archaeological and Ethnographic Studies of Homegardens in the American Neotropics. *Journal of Anthropological Archaeology* 69:101469, p. 5.

10. VOPA used a soil classification system with five types, with Class I soils being the best and Class V the worst; see S. L. Fedick. 1996. An Interpretive Kaleidoscope: Alternative Perspectives on Ancient Agricultural Landscapes of the Maya Lowlands. In *The Managed Mosaic: Ancient Maya Agriculture and Resource Use*, edited by S. L. Fedick, pp. 107–131. University of Utah Press, Salt Lake City.

11. E. C. Wells and L. D. Mihok. 2010. Ancient Maya Perceptions of Soil, Land, and Earth. In *Soil and Cultures*, edited by E. R. Landa and C. Feller, pp. 311–327. Springer Press, New York.

12. R. Redfield and A. Villa Rojas. 1964 [1934]. *Chan Kom: A Maya Village*. Abridged ed. University of Chicago Press, Chicago, p. 82.

13. P. Cunningham-Smith and K. Emery. 2020. Dogs and People: Exploring the Human-Dog Connection. *Journal of Ethnobiology* 40:409–413. https://journals.sagepub.com/doi/10.2993/0278-0771-40.4.409.

14. Kimmerer, *Braiding Sweetgrass*, p. 139.

15. E. N. Anderson with A. Dzib Xihum de Cen, F. Medina Tzuc, and Pastor Valdez Chale. 2005. *Political Ecology in a Yucatec Maya Community*. University of Arizona Press, Tucson, pp. 116, 118–119.

16. Á. Past with X. Guzmán Bakbolom and X. Ernandes. 2005. *Incantations: Song, Spells and Images by Mayan Women*. Cinco Puntos Press, El Paso, TX, p. 111.

17. Redfield and Villa Rojas, *Chan Kom*, p. 139.

18. C. Vidal-Lorenzo and P. Horcajada-Campos. 2020. Water Rituals and Offerings to the Maya Rain Divinities. *European Journal of Science and Theology* 16:111–123.

19. Since most of the materials used in traditional Cha'a Cháak ceremonies are organic, their remains would leave little if any evidence in the archaeological record. But as seen at Cara Blanca Pool 1 and elsewhere (see Chapter 7), the archaeological record is still replete with other types of rituals that demonstrate the vital importance of engaging with gods, ancestors, and nonhumans.

20. For updates, see https://www.icj-cij.org/case/177.

Chapter 7

1. E. Z. Vogt. 1969. *Zinacantan: A Maya Community in the Highlands of Chiapas*. Belknap Press of Harvard University Press, Cambridge, MA, p. 71.

2. Á. Past with X. Guzmán Bakbolom and X. Ernandes. 2005. *Incantations: Song, Spells and Images by Mayan Women*. Cinco Puntos Press, El Paso, TX, pp. 138–139.

3. E. Z. Vogt. 1993. *Tortillas for the Gods: A Symbolic Analysis of Zinacanteco Rituals*. University of Oklahoma Press, Norman, p. 52.

4. Vogt, *Tortillas for the Gods*, p. 54.

5. Past, Guzmán Bakbolom, and Ernandes, *Incantations*, p. 50.

6. R. E. Reina and R. M. Hill, II. 1978. *The Traditional Pottery of Guatemala*. University of Texas Press, Austin, TX, pp. 232–233.

7. R. J. McGee. 1998. The Lacandon Incense Burner Renewal Ceremony: Termination and Dedication Ritual among the Contemporary Maya. In *The Sowing and the Dawning: Termination, Dedication, and Transformation in the Archaeological and Ethnographic Record of Mesoamerica*, edited by S. B. Mock, pp. 41–46. University of New Mexico Press, Albuquerque, p. 41.

8. K. E. Seligson and M. Chi Nah. 2020. Mul Meyaj Tía U Betá Jump'el Kaj: Working Together to Build a Community in Puuc Archaeology. *Heritage* 3:342–363, p. 350.

9. L. J. Lucero. 1994. *Household and Community Integration among Hinterland Elites and Commoners: Maya Residential Ceramic Assemblages of the Belize River Area*. Ph.D. dissertation, UCLA. University Microfilms, Ann Arbor, MI, Appendices C and D; see also A. Ford and L. J. Lucero. 2001. The Malevolent Demons of Ceramic Production: Where Have All the Failures Gone? *Estudios de Cultura Maya* 21:57–74.

10. For more examples and discussion, see L. J. Lucero. 2006. *Water and Ritual: The Rise and Fall of Classic Maya Rulers*. University of Texas Press, Austin, pp. 67–113; L. J. Lucero. 2010. Materialized Cosmology among Ancient Maya Commoners. *Journal of Social Archaeology* 10:138–167; and L. J. Lucero. 2008. Memorializing Place among Classic Maya Commoners. In *Memory Work: Archaeologies of Material Practices*, edited by B. J. Mills and W. H. Walker, pp. 187–205. School for Advanced Research Press, Santa Fe, NM.

11. Past, Guzmán Bakbolom, and Ernandes, *Incantations*, p. 74.

12. D. Z. Chase and A. F. Chase. 2011. Ghosts Amid the Ruins: Analyzing Relationships Between the Living and the Dead among the Ancient Maya at Caracol, Belize. In *Between the Dead and the Living in Mesoamerica*, edited by I. Shimada and J. Fitzsimmons, pp. 78–101. University of Arizona Press, Tucson.

13. A. J. Christenson. 2016. *The Burden of the Ancients: Maya Ceremonies of World Renewal from the Pre-Columbian Period to the Present.* University of Texas Press, Austin, p. 27.

14. W. R. Coe. 1990. *Excavations in the Great Plaza, North Terrace and North Acropolis of Tikal.* Tikal Report no. 14. University Museum, University of Pennsylvania, Philadelphia, pp. 525–554.

Chapter 8

1. Most of the comparison between Angkor and Maya cities is from L. J. Lucero, R. Fletcher, and R. Coningham. 2015. From "Collapse" to Urban Diaspora: The Transformation of Low-Density, Dispersed Agrarian Urbanism. *Antiquity* 89:1139–1154.

2. M. D. Coe and D. Evans. 2018. *Angkor and the Khmer Civilization.* 2nd ed. Thames and Hudson, London, pp. 241, 258.

3. J. S. Lansing. 1991. *Priests and Programmers: Technologies of Power in the Engineered Landscape of Bali.* University of Princeton Press, Princeton, NJ, pp. 6, 27, 44–52, 107–110.

4. A. K. Goel. 1999. Lotus: Food For the Body. *Hinduism Today.* https://www.hinduismtoday.com/magazine/july-1999/1999-07-lotus-food-for-the-body/.

5. D. Tian. 2008. *Container Production and Post-harvest Handling of Lotus (Nelumbo) and Micropropagation of Herbaceous Peony (Paeonia).* Ph.D. dissertation. Auburn University, Alabama, pp. 1, 12–13, 20, 27.

6. C. Geertz. 1980. *Negara: The Theatre State in Nineteenth-Century Bali.* Princeton University Press, Princeton, NJ, pp. 105, 131.

7. A. Ford. 1996. Critical Resource Control and the Rise of the Classic Period Maya. In *The Managed Mosaic: Ancient Maya Agriculture and Resource Use,* edited by S. L. Fedick, pp. 297–303. University of Utah Press, Salt Lake City.

8. A. J. McDonald and B. Stross. 2012. Water Lily and Cosmic Serpent: Equivalent Conduits of the Maya Spirit Realm. *Journal of Ethnobiology* 32: 73–106, p. 87.

9. S. Houston and D. Stuart. 1996. Of Gods, Glyphs and Kings: Divinity and Rulership among the Classic Maya. *Antiquity* 70:289–312, p. 299.

10. P. Swindels. 1983. *Waterlilies.* Croom Helm, Kent, UK, pp. 12–13.

11. See M. Dobkin de Rios. 1974. The Influence of Psychotropic Flora and Fauna on Maya Religion. *Current Anthropology* 15:147–164; W. A. Emboden. 1982. The Mushroom and the Water Lily: Literary and Pictorial Evidence for *Nymphaea* as a Ritual Psychotogen in Mesoamerica. *Journal of Ethnopharmacology* 5:139–148; and R. E. Schultes and A. Hoffman. 1992 [1979]. *Plants of the Gods: Their Sacred, Healing and Hallucinogenic Powers.* Healing Arts Press Rochester, VT.

12. D. L. Spess. 2008. *Soma: The Divine Hallucinogen.* Park Street, Rochester, VT, pp. 33, 35.

13. Lucero, Fletcher, and Coningham, From "Collapse" to Urban Diaspora.

14. M. Medina-Elizalde, S. J. Burns, D. W. Lea, Y. Asmerom, L. von Gunten, V. Polyak, M. Vuille, and A. Karmalkar. 2010. High Resolution Stalagmite Climate Record from the Yucatán Peninsula Spanning the Maya Terminal Classic Period. *Earth and Planetary Science Letters* 298:255–262.

15. Z. A. Nissen, K. C. Grauer, G. Dziki, and H. Hoover. 2023. Situating Households within an Urban Community: Recent Research at Aventura, an Ancient Maya City. *Research Reports in Belizean Archaeology* 18:257–264.

16. Lucero, Fletcher, and Coningham, From "Collapse" to Urban Diaspora.

Chapter 9

1. A. M. Tozzer. 1941. *Landa's Relación de los Cosas de Yucatán.* Papers of the Peabody Museum of American Archaeology and Ethnology, no. 28. Harvard University, Cambridge, MA, p. 87.

2. R. Fletcher. 2009. Low-Density, Agrarian-Based Urbanism: A Comparative View. *Insights* (Institute of Advanced Study, Durham University) 2(4):1–19.

3. P. Roberts. 2021. *Jungle: How Tropical Forests Shaped the World—and Us.* Basic Books, New York, p. 241.

4. For field reports and maps, see the Caracol project website: https://caracol.org/.

5. F. Estrada-Belli and A. Tokovinine. 2022. Chochkitam: A New Classic Maya Dynasty and the Rise of the Kaanu'l (Snake) Kindgom. *Latin American Antiquity* 33:713–732. https://doi.org/10.1017/laq.2022.43.

6. For a discussion of Maya warfare, see the section on Maya warfare in the 2023 issue of *Ancient Mesoamerica*: 34(1). https://www.cambridge.org/core/journals/ancient-mesoamerica/issue/B4EB67548A60816E4F17017AC1CE029A.

7. For example, see D. Wahl, L. Anderson, F. Estrada-Belli, and A. Tokovinine. 2019. Palaeoenvironmental, Epigraphic and Archaeological Evidence of Total Warfare among the Classic Maya. *Nature Human Behaviour* 3:1049–1054. https://www.nature.com/articles/s41562-019-0671-x; for a discussion of warfare in the inscriptions, see S. Martin. 2020. *Ancient Maya Politics: A Political Anthropology of the Classic Period 150–900 CE.* University of Cambridge Press, Cambridge, MA, pp. 196–236.

8. A. J. Christenson, translator. 2022. *The Title of Totonicapán.* University Press of Colorado, Louisville, pp. 107–108. This book was completed in 1554 by K'iche' Maya 30 years after the Spanish conquered western Guatemala.

9. For example, A. K. Scherer, C. Golden, S. Houston, M. E. Matsumoto, O. A. Alcover Firpi, W. Schroder, A. Roche Recinos, et al. 2022. Chronology and the Evidence for War in the Ancient Maya Kingdom of Piedras Negras. *Journal of Anthropological Archaeology* 6:101408. https://doi.org/10.1016/j.jaa.2022.101408.

10. K. Bassie-Sweet. 2021. *Maya Gods of War.* University Press of Colorado, Louisville.

11. See J. G. Fox. 1996. Playing with Power: Ballcourts and Political Ritual in Southern Mesoamerica. *Current Anthropology* 37:483–509; M. M. Stoll. 2023. Es Nuestra Tradicíon: The Archaeological Implications of an Ethnography on a Modern Ballgame in Oaxaca, Mexico. *Ancient Mesoamerica* 1–25. https://doi.org/10.1017/S0956536123000019; and M. Zender. 2004. Sport, Spectacle and Political Theater: New Views of the Classic Maya Ballgame. *PARI Journal* 4(4):10–12.

12. S. Martin and N. Grube. 2008. *Chronicle of the Maya Kings and Queens.* 2nd ed. Thames and Hudson, New York, p. 110.

13. J. E. Thompson. 1939. *Excavations at San Jose, British Honduras.* Carnegie Institution of Washington Publication no. 506. Carnegie Institution, Washington, DC.

14. See UNESCO World Heritage tentative list; https://whc.unesco.org/en/tentativelists/6623/; for articles about Naranjo Sa'aal, see https://www.mesoweb.com/CATNYN/index.html.

Chapter 10

1. E. N. Anderson with A. Dzib Xihum de Cen, F. Medina Tzuc, and Pastor Valdez Chale. 2005. *Political Ecology in a Yucatec Maya Community.* University of Arizona Press, Tucson, p. 167.

2. See their website: https://www.umass.edu/gateway/research/indigenous-knowledges?et_rid=49183883&et_cid=4960921.

3. D. I. Kertzer. 1988. *Ritual, Politics, and Power*. Yale University Press, New Haven, CT, p. 67.

4. D. B. Lee. 2000. *Old Order Mennonites: Rituals, Beliefs, and Community*. Burnham Publishers, Chicago, p. 5.

5. J. S. Lansing. 1991. *Priests and Programmers: Technologies of Power in the Engineered Landscape of Bali*. University of Princeton Press, Princeton, NJ, pp. 111–126.

6. A. Kennedy-Shaffer. 2009. *The Obama Revolution*. Phoenix Books, Beverly Hills, CA, p. 61.

7. C. L. Hogue. 1993. *Latin American Insects and Entomology*. University of California Press, Berkeley, p. 72.

8. R. W. Kimmerer. 2013. *Braiding Sweetgrass: Indigenous Wisdom, Scientific Knowledge, and the Teachings of Plants*. Milkweed Editions, Minneapolis, MN, p. 285.

9. C. T. Darimont, R. Cooke, M. L. Bourbonnais, H. M. Bryan, S. M. Carlson, J. A. Estes, M. Galetti, et al. 2023. Humanity's Diverse Predatory Niche and its Ecological Consequences. *Communications Biology* 6:609, p. 1. https://doi.org/10.1038/s42003-023-04940-w.

10. Joint Secretarial Order on Fulfilling the Trust Responsibility to Indian Tribes in the Stewardship of Federal Lands and Waters. https://www.usda.gov/sites/default/files/documents/joint-so-3403-stewardship-tribal-nations.pdf.

11. V. Watts. 2013. Indigenous Place-Thought and Agency amongst Humans and Nonhumans (First Woman and Sky Woman go on a European World Tour!). *Decolonization: Indigeneity, Education and Society* 2:20–34, p. 27.

12. https://www.liveabout.com/facts-about-pools-spas-swimming-safety-2737127

13. For example, see H. S. Conard. 2017 [1909]. *Water Lilies and How to Grow Them: Including the Proper Making of Ponds*. Forgotten Books, London.

14. See https://www.linkedin.com/in/dalayna-williams-75560724/.

15. https://www.un.org/sustainabledevelopment/water-and-sanitation/

16. United Nations. 2023. *The United Nations World Water Development Report 2023: Partnerships and Cooperation for Water*. UNESCO, Paris, p. 2. https://www.unesco.org/reports/wwdr/2023/en.

17. E. Stukenbrock and S. Gurr. 2023. Address the Growing Urgency of Fungal Disease in Crops. Comment. *Nature* 617:31–34. https://www.nature.com/articles/d41586-023-01465-4.

18. J. Mt. Pleasant. 2015. A New Paradigm for Pre-Columbian Agriculture in North America. *Early American Studies: An Interdisciplinary Journal* 13:374–412.

19. Sustainable intensification methods were assessed in 286 projects in 57 countries (study 1) and 40 projects in 20 African countries (study 2). "In both, several million farmers on tens of megahectares had adopted practices that had led to yield increases of 79% (study 1) and 113% (study 2)" over a period of 3 to 10 years. In these instances, farmers voluntarily adopted the strategies. J. Pretty. 2018. Intensification for Redesigned and Sustainable Agricultural Systems. *Science* 362:eaav0294, p. 2.

20. "Corn and soy grown in more complex rotations exhibited greater yields and more stability during hot and dry periods in the USA... and water infiltration that reduced drought effects was markedly improved in complex organic rotations compared to conventional monocultures." C. Kremen and A. M. Merenlender. 2018. Landscapes that Work for Biodiversity and People. *Science* 362:eaau6020, Table S2.

21. C. Robin. 2013. *Everyday Life Matters: Maya Farmers at Chan*. University Press of Florida, Gainesville, pp. 120–123.

22. M. J. Schmidt, S. L. Goldberg, M. Heckenberger, C. Fausto, B. Franchetto, J. Watling, Helena Lima, et al. 2023. Intentional Creation of Carbon-Rich Dark Earth Soils in the Amazon. *Science Advances* 9:eadh8499.

23. A. M. Thiel and M. B. Quinlan. 2022. Homegarden Variation and Medicinal Plant Sharing among the Q'eqchi' Maya of Guatemala. *Economic Botany* 76:16–33. https://doi.org/10.1007/s12231-021-09537-0.

24. F. Keck. 2019. Livestock Revolution and Ghostly Apparitions: South China as a Sentinel Territory for Influenza Pandemics. *Current Anthropology* 60(S20):S183–S353, p. S183.

25. M. Worobey, J. I. Levy, L. M. Serrano, A. Crits-Christoph, J. E. Pekar, S. A. Goldstein, A. L. Rasmussen, et al. 2022. The Huanan Seafood Wholesale Market in Wuhan was the Early Epicenter of the COVID-19 Pandemic. *Science* 377:951–959. https://doi.org/10.1126/science.abp8715.

26. L. Greenspoon, E. Krieger, R. Sender, and R. Milo. 2023. The Global Biomass of Wild Mammals. *Proceedings of the National Academy of Sciences* 120:e2204892120. https://doi.org/10.1073/pnas.2204892120.

27. https://worldanimalfoundation.org/dogs/how-many-dogs-are-in-the-world/; https://worldanimalfoundation.org/advocate/cat-statistics/

28. https://abcbirds.org/program/cats-indoors/cats-and-birds/; C. A. Lepczyk, J. E. Fantle-Lepczyk, K. D. Dunham, E. Bonnaud, J. Lindner, T. S. Doherty, and J. C. Z. Woinarski. 2023. A Global Synthesis and Assessment of Free-Ranging Domestic Cat Diet. *Nature Communication* 14:7809. https://doi.org/10.1038/s41467-023-42766-6.

29. O. J. Schmitz, M. Sylvén, T. B. Atwood, E. S. Bakker, F. Berzaghi, J. F. Brodie, J. P. G. M. Cromsigt, et al. 2023. Trophic Rewilding can Expand Natural Climate Solutions. *Nature Climate Change* 13:324–333. https://doi.org/10.1038/s41558-023-01631-6.

30. P. Roberts. 2021. *Jungle: How Tropical Forests Shaped the World—and Us*. Basic Books, New York, p. 155.

31. C. Halperin. 2021. Ancient Recycling: Considerations of the Wasteful, Meaningful, and Practical from the Maya Site of Ucanal, Peten, Guatemala. *Journal of Archaeological Method and Theory* 28:766–792. https://doi.org/10.1007/s10816-020-09490-7.

32. M. Deal. 1998. *Pottery Ethnoarchaeology in the Central Maya Highlands*. University of Utah Press, Salt Lake City.

33. L. M. Seymour, J. Maragh, P. Sabatini, M. Di Tommaso, J. C. Weaver, and A. Masic. 2023. Hot Mixing: Mechanistic Insights into the Durability of Ancient Roman Concrete. *Science Advances* 9:eadd1602. https://doi.org/10.1126/sciadv.add1602; T. Puiu. 2022. How the Ancient Romans Built Roads to Last Thousands of Years. *ZME Science*, Jan. 30, 2022. https://www.zmescience.com/science/how-roman-roads-were-built/.

34. C. Rodriguez-Navarro, L. Monasterio-Guillot, M. Burgos-Ruiz, E. Ruiz-Agudo, and K. Elert. 2023. Unveiling the Secret of Ancient Maya Masons: Biomimetic Lime Plasters with Plant Extracts. *Science Advances* 9:eadf6138.

35. M. L. Smith. 2019. *Cities: The First 6,000 Years*. Viking, New York, p. 180.

36. J. T. Murphy and C. L. Crumley (editors). 2022. *If the Past Teaches, What Does the Future Learn? Ancient Urban Regions and the Durable Future*. Delft University, Delft, the Netherlands, p. 83. https://bookrxiv.com/index.php/b/catalog/book/32.

37. Á. Past with X. Guzmán Bakbolom and X. Ernandes. 2005. *Incantations: Song, Spells and Images by Mayan Women*. Cinco Puntos Press, El Paso, TX, p. 179.

38. A. J. Christenson. 2007. *Popol Vuh: The Sacred Book of the Maya*. University of Oklahoma Press, Norman, pp. 126–127.

39. A. R. Wyatt. 2023. "An Instrument of Grace": Archaeological and Ethnographic Studies of Homegardens in the American Neotropics. *Journal of Anthropological Archaeology* 69:101469, p. 7.

40. N. C. Kawa. 2022. Ten Ways that Human Waste Isn't Waste at All. *Anthropology News* website, August 10, 2022. https://www.anthropology-news.org/articles/ten-ways-that-human-waste-isnt-waste-at-all; see also P. Annin. 2023. *Purified: How Recycled Sewage Is Transforming Our Water*. Island Press, Washington, DC.

41. Company slogan for Recompose (https://recompose.life/).

42. We also can have green weddings, birthday celebrations, gender reveals, and other rites of passage.

43. https://aquamationinfo.com/

44. Loop Biotech; https://loop-biotech.com/

45. https://returnhome.com/human-composting/

46. One of my students stated this in my 200-level "Climate Change and Civilization" course.

FURTHER READINGS

INTRODUCTION

B. M. Fagan and N. Durrani. 2022. *Archaeology: A Brief Introduction*. 13th ed. Routledge, New York.

T. R. Pauketat. 2023. *Gods of Thunder: How Climate Change, Travel, and Spirituality Reshaped Precolonial America*. Oxford University Press, New York.

M. Peuramaki-Brown and S. Batty. 2021. Belize Shows how Local Engagement is Key in Repatriating Cultural Artifacts from Abroad. *The Conversation*, November 14. https://theconversation.com/belize-shows-how-local-engagement-is-key-in-repatriating-cultural-artifacts-from-abroad-171363?utm_source=dlvr.it&utm_medium=facebook&fbclid=IwAR1u9SHMeXKmikZSe1yzn-6OpVDT1wtnpEdOV28UBQMtalIBea1ZHtn1r7w.

M. Saunders (editor). 2022. *Tales from the Field*. Precolumbia Mesoweb Press, San Francisco.

V. Strang. 2023. *Water Beings: From Nature Worship to the Environmental Crisis*. Reaktion Books, London.

Valley of Peace Archaeology Project website. https://publish.illinois.edu/valleyofpeace/ (for photos, videos, blogs, reports, publications and more).

CHAPTER 1

T. Ardren. 2023. *Everyday Life in the Classic Maya World*. Cambridge University Press, Cambridge, UK.

R. Argivo. 1994. *Sastun: My Apprenticeship with a Maya Healer*. Harper, San Francisco.

B. Goldfarb. 2023. *Crossings: How Road Ecology is Shaping the Future of Our Planet.* W. W. Norton, New York.

J. T. Larmon, L. J. Lucero, and F. Valdez, Jr. (editors). 2022. *Sustainability and Water Management in the Maya World and Beyond.* University Press of Colorado, Louisville.

Maya Ethnobotany website. http://www.maya-ethnobotany.org/.

P. Roberts. 2019. *Tropical Forest Prehistory, History, and Modernity.* Oxford University Press, London.

P. Roberts. 2021. *Jungle: How Tropical Forests Shaped the World—and Us.* Basic Books, New York.

V. Schlesinger. 2001. *Animals and Plants of the Ancient Maya: A Guide.* University of Texas Press, Austin.

CHAPTER 2

M. D. Coe and S. D. Houston. 2015. *The Maya.* 9th ed. Thames and Hudson, New York.

I. Cojtí Ren. 2021. The *Saqirik* (Dawn) and Foundation Rituals among the Ancient K'iche'an Peoples. In *The Myths of the Popol Vuh in Cosmology, Art and Ritual,* edited by H. Moyes, A. J. Christenson, and F. Sasche, pp. 77–90. University Press of Colorado, Louisville.

M. S. Findlay. 2020. *Cultural Traditions of Ancient Mesoamerica.* Cognella Academic Publishing, San Diego, CA.

C. A. Kray. 2023. *Maya-British Conflict at the Edge of the Yucatecan Caste War.* University Press of Colorado, Louisville.

J. T. Larmon, H. G. McDonald, S. Ambrose, L. R. G. DeSantis, and L. J. Lucero. 2019. A Year in the Life of a Giant Ground Sloth During the Last Glacial Maximum in Belize. *Science Advances* 5(2):eaau1200.

L. J. Lucero. 2006. *Water and Ritual: The Rise and Fall of Classic Maya Rulers.* University of Texas Press, Austin.

R. M. Palacios. 2022. Maya Literature. *Oxford Research Encyclopedia of Literature.* https://doi.org/10.1093/acrefore/9780190201098.013.1295.

J. Wainwright. 2021. The Maya and the Belizean State: 1997–2004. *Latin American and Caribbean Ethnic Studies* 17(3):320–349. https://doi.org/10.1080/17442222.2021.1935694.

CHAPTER 3

D. Finamore and S. D. Houston (editors). 2010. *Fiery Pool: The Maya and the Mythic Sea.* Peabody Essex Museum and Yale University Press, New Haven, CT.

L. J. Lucero. 2023. Ancient Maya Reservoirs, Constructed Wetlands, and Future Water Needs. *Proceedings of the National Academy of Sciences* 120:e2306870120. https://doi.org/10.1073/pnas.2306870120.

L. J. Lucero, J. D. Gunn, and V. L. Scarborough. 2011. Climate Change and Classic Maya Water Management. *Water* 3:479–494.

L. J. Lucero and J. T. Larmon. 2018. Climate Change, Mesoamerica and the Classic Maya Collapse. In *Climate Changes in the Holocene: Impacts and Human Adaptation*, edited by E. Chiotis, pp. 165–181. CRC Press, London.

V. L. Scarborough. 2003. *The Flow of Power: Ancient Water Systems and Landscapes*. School of American Research Press, Santa Fe, NM.

United Nations. 2023. *The United Nations World Water Development Report 2023: Partnerships and Cooperation for Water*. UNESCO, Paris. https://www.unesco.org/reports/wwdr/2023/en.

C. Vidal-Lorenzo and P. Horcajada-Campos. 2020. Water Rituals and Offerings to the Maya Rain Divinities. *European Journal of Science and Theology* 16(2):111–123.

CHAPTER 4

G. E. Chacon. 2018. *Indigenous Cosmolectics: Kab'awil and the Making of Maya and Zapotec Literatures*. University of North Carolina Press, Chapel Hill.

A. J. Christenson. 2007. *Popol Vuh: The Sacred Book of the Maya*. University of Oklahoma Press, Norman.

L. J. Lucero. 2018. A Cosmology of Conservation in the Ancient Maya World. *Journal of Anthropological Research* 74:327–359.

D. S. Stuart. 2011. *The Order of Days: The Maya World and the Truth about 2012*. Harmony Books, New York.

D. Suzuki and P. Knudts. 1993. *Wisdom of the Elders: Sacred Native Stories of Nature*. Bantam Books, New York.

CHAPTER 5

M. A. Astor-Aguilera. 2010. *The Maya World of Communicating Objects: Quadripartite Crosses, Trees, and Stones*. University of New Mexico Press, Albuquerque, NM.

R. W. Kimmerer. 2013. *Braiding Sweetgrass: Indigenous Wisdom, Scientific Knowledge, and the Teachings of Plants*. Milkweed Editions, Minneapolis, MN.

L. J. Lucero and A. Kinkella. 2015. Pilgrimage to the Edge of the Watery Underworld: An Ancient Maya Water Temple at Cara Blanca, Belize. *Cambridge Archaeological Journal* 25:163–185.

L. J. Lucero, J. Harrison, J. Larmon, Z. Nissen, and E. Benson. 2016. Prolonged Droughts, Short-Term Responses and Diaspora: The Power of Water and Pilgrimage at the Sacred Cenotes of Cara Blanca, Belize. *WIREs Water*. https://doi.org/10.1002/wat2.1148.

Á. Past with X. Guzmán Bakbolom, and X. Ernandes. 2005. *Incantations: Song, Spells and Images by Mayan Women*. Cinco Puntos Press, El Paso, TX.

CHAPTER 6

E. N. Anderson with A. Dzib Xihum de Cen, F. Medina Tzuc, and Pastor Valdez Chale. 2005. *Political Ecology in a Yucatec Maya Community*. University of Arizona Press, Tucson, AZ.

C. Cagnato. 2019. Prehistoric and Traditional Agriculture in Lowland Mesoamerica. *Oxford Research Encyclopedia of Environmental Science*. https://doi.org/10.1093/acrefore/9780199389414.013.174.

A. J. Christenson. 2016. *The Burden of the Ancients: Maya Ceremonies of World Renewal from the Pre-Columbian Period to the Present*. University of Texas Press, Austin, TX.

A. Ford and R. Nigh. 2015. *The Maya Forest Garden: Eight Millennia of Sustainable Cultivation of the Tropical Woodlands*. Left Coast Press, Walnut Creek, CA.

J. Mt. Pleasant. 2015. A New Paradigm for Pre-Columbian Agriculture in North America. *Early American Studies: An Interdisciplinary Journal* 13:374–412.

J. W. Palka. 2023. Ancestral Maya Domesticated Waterscapes, Ecological Aquaculture, and Integrated Subsistence. *Ancient Mesoamerica* 35(1):1–29. https://doi.org/10.1017/S0956536122000402.

A. R. Wyatt. 2023. "An Instrument of Grace": Archaeological and Ethnographic Studies of Homegardens in the American Neotropics. *Journal of Anthropological Archaeology* 69:101469. https://doi.org/10.1016/j.jaa.2022.101469.

CHAPTER 7

S. L. Fedick and L. S. Santiago. 2022. Large Variation in Availability of Maya Food Plant Sources During Ancient Droughts. *Proceedings of the National Academy of Sciences* 119(1):e2115657118. https://doi.org/10.1073/pnas.2115657118.

L. J. Lucero. 2010. Materialized Cosmology among Ancient Maya Commoners. *Journal of Social Archaeology* 10:138–167.

P. A. McAnany. 1995. *Living with the Ancestors: Kinship and Kingship in Ancient Maya Society*. University of Texas Press, Austin, TX.

S. B. Mock (editor). 1998. *The Sowing and the Dawning: Termination, Dedication, and Transformation in the Archaeological and Ethnographic Record of Mesoamerica*. University of New Mexico Press, Albuquerque, NM.

C. Robin. 2013. *Everyday Life Matters: Maya Farmers at Chan*. University Press of Florida, Gainesville, FL.

E. Z. Vogt. 1969. *Zinacantan: A Maya Community in the Highlands of Chiapas*. Belknap Press of Harvard University Press, Cambridge, MA.

CHAPTER 8

M. D. Coe and D. Evans. 2018. *Angkor and the Khmer Civilization*. 2nd ed. Thames and Hudson, London.

J. L. Fitzsimmons. 2009. *Death and the Classic Maya Kings*. University of Texas Press, Austin, TX.

L. J. Lucero. 2006. *Water and Ritual: The Rise and Fall of Classic Maya Rulers*. University of Texas Press, Austin, TX.

L. J. Lucero, R. Fletcher, and R. Coningham. 2015. From "Collapse" to Urban Diaspora: The Transformation of Low-Density, Dispersed Agrarian Urbanism. *Antiquity* 89:1139–1154.

S. Martin. 2020. *Ancient Maya Politics: A Political Anthropology of the Classic Period 150–900 CE*. University of Cambridge Press, Cambridge, UK.

L. Schele and D. Freidel. 1990. *A Forest of Kings: The Untold Story of the Ancient Maya*. William Morrow, New York.

L. Schele and M. E. Miller. 1986. *The Blood of Kings: Dynasty and Ritual in Maya Art*. George Braziller, New York.

CHAPTER 9

A. S. Z. Chase, A. F. Chase, and D. Z. Chase (editors). 2024. *Ancient Mesoamerican Population History: Urbanism, Social Complexity, and Change*. University of Arizona Press, Tucson, AZ.

D. Z. Chase, J. Lobo, G. M. Feinman, D. M. Carballo, A. F. Chase, A. S. Z. Chase, S. R. Hutson, et al. 2023. Mesoamerican Urbanism Revisited: Environmental Change, Adaptation, Resilience, Persistence, and Collapse. *Proceedings of the National Academy of Sciences* 120:e2211558120.

S. D. Houston, C. Brittenham, C. Mesick, A. Tokovinine, and C. Warinner. 2009. *Veiled Brightness: A History of Ancient Maya Color*. University of Texas Press, Austin, TX.

S. Hutson and T. Ardren (editors). 2020. *The Maya World*. Routledge, London.
M. L. Smith. 2019. *Cities: The First 6,000 Years*. Viking, New York.

CHAPTER 10

C. Barash (editor). 2001. *Water Gardens: How to Create Beautiful Fountains, Ponds, and Streams*. Better Homes & Gardens. Des Moines, IA.

R. Coningham and L. J. Lucero. 2021. Urban Infrastructure, Climate Change, Disaster and Risk: Lessons from the Past for the Future. *Journal of the British Academy* 9(s8):79–114.

M. Dedrick, P. A. McAnany, and A. I. Batún-Alpuche. 2023. Tracing the Structural Consequences of Colonialism in Rural Yucatán, Mexico. *American Anthropologist* 125:390–403. https://doi.org/10.1111/aman.13843.

B. M. Fagan and N. Durrani. 2021. *Climate Chaos: Lessons on Survival from Our Ancestors*. PublicAffairs, New York.

S. Fiske, S. Crate, C. Crumley, K. Galvin, H. Lazarus, G. Luber, L. Lucero, et al. 2015. *Changing the Atmosphere: Anthropology and Climate Change*. American Anthropological Association Climate Change Task Force Report, Arlington, VA. https://doi.org/10.13140/RG.2.1.1971.6328.

E. Graham. 2011. *Maya Christians and their Churches in Sixteenth-Century Belize*. University Press of Florida, Gainesville, FL.

J. S. Lansing. 1991. *Priests and Programmers: Technologies of Power in the Engineered Landscape of Bali*. University of Princeton Press, Princeton, NJ.

P. A. McAnany. 2016. *Maya Cultural Heritage: How Archaeologists and Indigenous Communities Engage the Past*. Rowman & Littlefield, Lanham, MD.

R. Menchú. 1983. *I, Rigoberta Menchú: An Indian Woman in Guatemala*. Edited by E. Burgos-Debray. Verso Press, New York.

V. D. Montejo. 2005. *Maya Intellectual Renaissance: Identity, Representation, and Leadership*. University of Texas Press, Austin, TX.

J. T. Murphy and C. L. Crumley (editors). 2022. *If the Past Teaches, What Does the Future Learn? Ancient Urban Regions and the Durable Future*. Delft University, Delft, Netherlands.

M. Nelson. 2013. *The Wastewater Gardener: Preserving the Planet One Flush at a Time*. Synergetic Press, Santa Fe, NM.

G. Thunberg. 2023. *The Climate Book: The Facts and the Solutions*. Penguin Press, New York.

UN-Water. 2019. *Water and Climate Change—UN-Water Policy Brief*. UN-Water, Geneva. https://www.unwater.org/sites/default/files/app/uploads/2019/10/UN_Water_PolicyBrief_ClimateChange_Water.pdf.

L. Zeldovich. 2021. *The Other Dark Matter: The Science and Business of Turning Waste into Wealth and Health*. University of Chicago Press, Chicago.

INDEX

adaptation, 3–4, 153
agriculture
   extensive, slash-and-burn, fallow, 27–29, 108, 110, 160, 183, 188–189
   home gardens, 2, 19, 87, 107–108, 111, 163, 177, 187–188
   intensive, 108, 160, 188–189
   milpas (fields), 19, 42–43, 73, 87, 107–113, 115–116, 176, 187–188, 198
   mollisols, 34, 39, 111–112
   planting, 83–84, 115, 117, 119, 184, 187–189
   raised fields, 35–37, 108
   soil fertility and distribution, 110–111, 160, 166–167, 192
   terraces, 35–37, 108, 112, 118–119, 176, 188, 193–194, 201
   Three Sisters, 106, 115, 188
Angkor, 145–149, 160

bajos (seasonal wetlands or swamps), 34, 38–40, 51–52, 108, 110–111, 152–153, 183

Barton Ramie, xi–xii, 43, 131, 171–172
Belize
   Institute of Archaeology, xiii, xvi, 27, 29, 52, 98, 158–159
   relations with Guatemala, 119–120
bottom-up approaches, 3, 182, 202
burials. *See also* ceremonies
   bundle, 97–98, 133
   funerary offerings, 71–73, 79–81, 98, 125, 134

Calakmul, 33, 39–40, 54, 166–167, 169
calendrics and time
   calendar round, 72–73, 133–134
   cyclical, 42, 63–64, 67–68, 93–94, 153–154
   Haab (solar calendar), 72–73
   long count, 42, 153–154
   short count, 42
   Tzolk'in (ritual calendar), 72–73, 134

233

INDEX

Cara Blanca. *See also* pilgrimage
  ceremonial buildings, 92–93
  ceremonies at *cenotes*, 89, 92–93, 95–97, 99
  human caches, 97–98, 133
  lakes, 9, 29, 41, 90–92, 103
  pilgrimage destination, 9, 76, 83–84, 90, 103–104, 183–184
  sinkholes or *cenotes*, 53, 89–92, 95, 98–104
  underwater (diving, megafauna fossils, giant sloth), 30–31, 100–104
Caracol, 33, 39–40, 54, 112, 133–134, 161–162, 166
cardinal directions, 65, 68, 70, 118, 133, 139–140, 142–143
ceiba, tree of life, 18, 63, 75–76, 86
ceramic production, 126–131, 180
ceremonies. *See also ch'e'n*
  ancestors, 64, 69–70, 72–74, 89, 110, 113, 118, 142–143, 150–151, 179–180
  animation, 69–70, 124–126, 129, 134, 161, 181–182
  Cha'a Chaak (rain ceremony), 47, 89, 92–93, 96, 117–118, 161, 186
  de-animation, 93, 96–99, 140–143, 181–182
  feasts, feasting, 38, 95–96, 141–142, 146, 152
  funerary, 71–74, 80–81, 125, 142–143, 161
  planting and harvesting, 113, 117–118, 150–151
  renewal, *see* renewal
  ritual deposits, 69–70, 73, 122–123, 125, 131–133, 138–139, 143. *See also* burials

ch'e'n, 72, 75–77, 89–104, 118, 179–180. *See also* ceremonies, pilgrimage
  communication with gods, ancestors, 64, 72, 76, 89, 96
  dead, the, 77, 79, 81, 133
  watery, 76, 83–84, 88–89, 92–93, 96–97. *See also* Cara Blanca
ch'ulel (soul), 64, 66–67, 69–70, 73–74, 124–125, 177
k'exol (replacement), 67, 124–125
recycled, 67, 72, 81, 134, 198–199
cities and urbanism
  gardens, milpas, orchards, 57–59, 158–159, 163, 166, 177, 190, 192–193
  green cities, 2–4, 183, 192–193
  large plazas, 38, 58–59, 73–74, 146, 152–153, 155–156, 166, 181–182, 196
  markets, 40, 160–163, 166, 192
  monumental architecture, 38, 40, 54, 60, 63–64, 147–149, 165, 171–174, 192–193
  sak b'eh (causeway), 39–40, 89, 161–162, 166, 195–196
  urban planning, layout, 58–60, 159, 166, 186, 192–193
collaboration
  direct or forest collaboration, 14, 19, 63, 87–88, 109, 176–178, 189
  indirect, *see* pilgrimage
constructed wetlands (CWs), 48, 56–57, 186–187, 193, 199
  swimming pools, x, 186

death begets life, destruction begets creation, *see* renewal
deforestation

INDEX

atmosphere, ix, 5, 17, 160
CO_2, 17, 22
erosion, 9, 17, 59, 153–154
   impact of plowing, xi–xii, 9, 26, 107–108, 182
domesticated foods, 19, 35–37, 106–107, 113–115, 187–188
Dos Pilas, 73, 167
droughts, 154. See also seasonality
   annual, 12–13, 15–16, 40, 53–55, 150–151
   crop failure, famine, 40, 77–79, 145, 154
   prolonged, severe, 8–9, 14, 29, 40, 60–61, 103, 153, 179, 201
dualism, complementary opposition, 5–6, 32–33, 66, 176

everything is connected, 68–69, 72, 106, 113, 123, 202–203

family histories, 27–29, 69–70, 109–110, 122, 124–125, 131–143
farming, see agriculture
forest entities, products, 40, 88–89, 159
   construction materials, 2, 40, 66–67, 83–84, 87, 144–145, 160–161, 185, 192
   foodstuffs (berries, game, nuts, fruit), 21, 35–37, 84, 86–88, 110, 183
   fuel, 2, 21, 40, 83–84, 87, 160–161
   hardwoods, 14–16, 27, 86–88, 168, 183
   medicinal, 19, 40, 57, 66–67, 87–88, 160–161, 182–183, 186

fossil fuels, greenhouse gases, see unsustainable practices

global climate change, 7, 17, 48–49, 51–52, 172–173, 202

house mounds, 53, 69–70, 111–112. See also mounds
household
   building blocks of society, 3, 202–203
   ceremonies, rituals, 124–125, 133, 139, 179–182. See also ceremonies
   items, xiii, 12–13, 19, 63–64, 73–74, 122, 182–183
   production, 127–128, 131, 161, 163
   unit of action, 3, 176, 178, 202–203

iconography
   fauna and flora, 18, 21–22, 24
   water and kings, 57–58, 60, 147–149
Indigenous. See also worldviews
   knowledge and practices, 3–4, 178–179, 188–189, 204
   peoples, xiii, 6, 43–45, 178, 185–186, 200–201
   places and plants, 44, 107–108

kingship, xi, 147, 160
   demise, collapse, 2, 7–8, 30–31, 152–157, 162
   no kings, 131, 155–157
   royal titles, 38–39, 147, 167
   water managers, 12–13, 40, 153

INDEX

limestone
  bedrock, 15–17, 48, 60–61, 75–76, 79, 100, 198
  cut limestone, 52, 95, 100, 122, 139–140, 195
  karstic landscape, 56, 74, 83
  plaster manufacture, 48–49, 54, 58–59, 84, 151, 183, 196
  porous, 15–16, 48, 56, 60–61, 75–76, 108–109, 198
  temper for ceramics, 84, 123, 127–130, 183, 196
logging, 27–29, 171–172, 184
looting, 7, 12–13, 29, 94, 159, 166, 169–174

Maya collapse, *see* kingship
Maya gods
  Chahk, 38, 68, 76, 90, 110, 118, 149–150, 179–180, 186
  Death Gods, 68–69, 90
  Earth God(s), 78, 125–126, 128–129, 180
  K'inich Ajaw, 38, 168, 179–180
  Maize God, 22, 68–69, 73–74, 106, 110, 152
Maya hieroglyphs, inscriptions, 39–40, 63–64, 112–113, 133–134, 147, 153–154, 158, 167–169
Maya history
  First Americans and Archaic, 31–32, 43, 113–114, 156
  Preclassic and Early Classic, 59–60, 138–139
  Late Classic, 17–18, 29, 40, 60, 76, 109, 160, 166, 193–194
  Terminal Classic, 41–42, 49, 76, 92, 94–95, 119, 162, 179
  Postclassic, 29, 41–42, 74, 90, 98, 119
  colonial rule, 2, 42–43, 45, 119, 178–179
  civil wars, 1, 8, 12, 45–46, 120
  massacres and the missing, 45–46
Maya people
  commoners, 32–33, 121, 133–134, 144–145, 180
  elites, 38, 40, 135–138, 140–141, 144, 150, 156, 160
  foremen and excavation assistants, x, 1, 10–11, 94, 196
  royals, 32–33, 40, 133, 153, 160–162, 166. *See also* kingship
Maya witches and witchcraft persecution, 77–82, 98, 168
Mayan languages, ethnic groups
  K'iche', 43–45, 63–64, 72–73, 98, 105, 175
  Lacandon, 66–67, 79, 129, 199
  Mopan, 1, 10–12, 21
  Q'eqchi', 10–12, 21, 190
  Tzeltal, 46, 78, 127, 196
  Tzotzil, 46, 64, 66, 70, 72, 78, 89, 126
  Yucatec, 66–68, 84, 89, 98, 113, 115, 158
Mennonite agricultural practices, xi–xii, 26–29, 107–108, 111
migration, 118–120, 175, 200–202
mimicking forest biodiversity, 13, 19, 26, 107–108, 182–183
mounds, xi–xii, 9, 26–29, 52–53, 99, 111–112, 155

Naranjo Sa'aal, xiii, 39–40, 150, 166–167, 171–174

nonhuman kinship with humans, 2, 42, 65–66, 82, 175–176, 178
northern lowlands
 Chichén Itzá, 33, 41–42, 90, 101, 169
 K'uk'ulcan (feathered serpent), 41–42
 Postclassic florescence, 33, 41–42
 Puuc, 42, 54, 129
 versus the southern lowlands, 33–34

organic waste, 197–200
 contamination, 56, 60–61
 fertilizer, 57, 186, 192, 198–200
origin and creation histories, 5, 31–32, 169
 Hero Twins, 32–33, 68–69, 152
 Popol Vuh, 63–64, 74, 105, 152, 175

Palenque, xii, 18, 72, 158, 165, 198–199
pilgrimage. *See also* Cara Blanca
 ceremonial circuit, 38, 89, 92–100, 183–184
 conservation, 83–84, 103–104, 183–184
 indirect collaboration, interaction, 83–84, 88–89, 103–104, 183–184
political systems
 heterarchical, self-organizing, 139, 144, 159–160, 162–163
 hierarchical, 160–161, 163
 surplus, 145, 162–163
 tribute, see tribute

population size, density, distribution, 17–18, 32, 38, 40, 106–107, 109–110, 160
portals, *see ch'e'n*

renewal
 cyclical, 42, 63–64, 67–68, 93–94, 153–154
 death begets life, destruction begets creation, 14–15, 67–68, 73–74, 109, 200
 life histories, 72–74, 93, 103
 rebirth, 41–42, 71–72, 125
reservoirs
 aquatic biota, 56–58, 90, 186
 drinking water, 48–49, 54, 152–153, 166, 186
 self-cleaning, x–xii, 12–13, 55–62, 186–187. *See also* constructed wetlands
rituals. *See also* ceremonies
 promote change, transformation, 38, 144, 161, 179–182
 promote solidarity, 38, 144, 161, 179–182
 specialists (*curanderos*, healers, priests), 78, 96, 126, 150, 161, 169, 182
rural farmsteads, xii, 39–40, 42–44, 112, 163, 183, 192. *See also* agriculture

San Jose, 170
Saturday Creek
 SC2, 135–138
 SC3, 134–138, 141
 SC18, 134–135, 142–143
 SC78, 139–141
 SC85, 134–135, 142

seasonality
  dry season, droughts, *see* droughts
  floods, flooding, 48, 50–53, 167, 193, 197
  hurricanes, 14, 47, 51–53, 55, 68, 109, 158–159
  tropical storms, 15–16, 47, 49, 52–53, 68, 117, 159
  wet, rainy season, 15–16, 25, 39, 48–53, 108, 117, 160–162
  wildfires, 17, 52, 55, 109, 171–172
shell
  freshwater, 21, 71–72, 95–96, 142–143, 155–156
  marine, 21, 38, 71–72, 140–141, 167
Spanish invasion and impacts
  congregación, 1, 42–43
  epidemic diseases, 1, 8, 22–23, 43, 156
  forced conversion, missionization, 8–9, 44, 63, 203–204
  settler colonialism, 45
staple foods (maize, beans, squash, manioc), 32–33, 113, 146, 155–156, 161–162
subsistence. *See also* agriculture, collaboration
  cooking, 64, 84, 88, 122, 183
  hunting, 2, 66–67, 74, 103–104, 113–114, 126
sustainable practices
  diversity, 17, 26, 86, 113–114, 176–177, 188–190
  green cities, 2–4, 12–13, 183, 192–193
  green, renewable technology, 177, 194–195
  hunting, fishing, 185–186, 193, 196
  local networks, 3–4, 183, 192–193
  multicropping, diverse crops, 106–107, 177, 187–188
  reforest, parks, 184–186, 191–192, 197, 199–200, 202
  reuse, repurpose, 195–200
  traditional knowledge, 12, 46, 63, 176–178, 189

Tikal
  Chak Tok Ich'aak II, 149–152
  North Acropolis, 134–135, 138–139, 143
  reservoirs, *see* reservoirs
top-down approaches, 3, 144–145, 178–179, 182, 202
tribute (goods, services, labor), 8, 38, 139, 145–146, 151, 158, 160, 192
tropical environment
  biota, *see* forest entities, products
  endemic diseases, parasites, 2, 14, 22–23, 56–57, 103–104, 151
  heat, humidity, 7, 17, 49, 53–55, 108–109, 122, 145–146, 184
  rainfall dependent, 12–13, 47, 106–107, 186
  savanna, 22, 30–31, 52–53
  tropical belt, 7, 14, 48–49

unsustainable practices
  capitalism, 4–6
  chemical pesticides, herbicides, fertilizers, xi–xii, 14, 26–27, 103, 107–108, 115, 182, 190
  exploitation, 5–6, 112–113, 185

extinction, x, 2, 17–18, 31, 175–176, 185, 191
finite, nonrenewable resources, 175, 177, 194
fossil fuels, ix, 175, 177, 186–187, 192–193
monocropping, xi–xii, 14, 26, 108, 115–116, 182, 188, 203–204
overuse of entities, x, 5, 59–60, 73–74, 154, 180, 184
privileging humans, x, 2–4, 6, 177, 202–203
short term, xii, 157, 178, 194–195, 202–204
urban diaspora, 40–41, 54, 59–60, 103, 149, 154
urban–rural interaction, 12–13, 161–163, 192–193

Valley of Peace Archaeology (VOPA)
  Cara Blanca, *see* Cara Blanca
  salvage archaeology, 9, 29
  Saturday Creek, *see* Saturday Creek
  Valley of Peace Village, 12, 45–46, 49, 87
  Yalbac, *see* Yalbac
  Yalbac Ranch, *see* Yalbac Ranch

warfare, 167–169, 194
water
  aguadas, xii, 9, 55–56, 58–59, 109, 152–153
  aquifer, groundwater, 9, 30–31, 43, 90–92, 131, 155
  desiccate, dry up, evaporation, 15–17, 53, 58, 92, 160

evapotranspiration, 17, 160
reservoirs, *see* reservoirs
rivers and creeks, 51, 53, 108, 153
suhuy ha' (pure water), 89, 96
water management, xi, 58–59, 186
water quality, 17, 48–49, 56–57, 59–60, 154, 184, 187
water lilies (*Nymphaea ampla*) and clean water, 18, 57–58, 90, 147, 154
way, animal spirit companion, 32–33, 66–67, 79
wits (lineage mountains), 66, 72, 76, 80–81, 89, 101, 103, 168
world maintenance, balance
  coexistence, ix, 2–4, 32, 62, 64, 72, 175–177
  collaboration, *see* collaboration
  merged existence, 93–96, 100, 160–161, 180
  renewal, rebirth, *see* renewal
worldviews
  anthropocentric, x, 4, 32, 65–66, 202–203
  Christianity, Catholicism, 4–5, 8, 44, 64, 77, 126, 203–204
  nature vs. culture, 4–7, 65–66, 177, 200–201
  nonanthropocentric, 2–4, 7, 9, 12–13, 63–64
  Western, x, 4, 31–32, 107–108, 115, 177–179

Yalbac, 9, 27, 51–52, 71–72, 87, 103–104, 112, 158–159
Yalbac Ranch, 27, 51–52, 54–55, 94, 170, 184